Friedrich Ruttner

Biogeography and Taxonomy of Honeybees

With 161 Figures

Springer-Verlag
Berlin Heidelberg New York
London Paris Tokyo

Professor Dr. FRIEDRICH RUTTNER
Bodingbachstraße 16
A-3293 Lunz am See

Legend for cover motif: Four species of honeybees around the area of distribution.

ISBN 3-540-17781-7 Springer-Verlag Berlin Heidelberg New York
ISBN 0-387-17781-7 Springer-Verlag New York Berlin Heidelberg

Library of Congress Cataloging in Publication Data. Ruttner, Friedrich. Biogeography and taxonomy of honeybees/Friedrich Ruttner. p. cm. Bibliography: p. Includes. index. ISBN 0-387-17781-7 (U.S.) 1. Apis (Insects) 2. Honeybee. I. Title. QL568.A6R88 1987 595.79'9--dc19

The use of registered names, trademarks, etc. in this publication does not imply, even in th absence of a specific statement, that such names are exempt from the relevant protective laws and regulations and therefore free for general use.

Data conversion and bookbinding: Appl, Wemding.
Offsetprinting: aprinta, Wemding
2131/3130-543210

Preface

Honeybees are as small as flies or as large as hornets, nesting in narrow cavities of trees and rocks or in the open on large limbs of trees 30 m above ground. They occur in tropical zones and in the forests of the Ural mountains, they survive seven months of winter and even longer periods of drought and heat. Historically, they lived through a extended time of stagnation in the tropics from the mid-Tertiary, but then experienced an explosive evolution during the Pleistocene, resulting in the conquest of huge new territories and the origin of two dozen subspecies in *Apis mellifera*.

This vast geographic and ecologic diversification of the genus *Apis* was accompanied by a rich morphological variation, less on the level of species than at the lowest rank, the subspecies level. Variation being exclusively of a quantitative kind at this first step of speciation, traditional descriptive methods of systematics proved to be unsatisfactory, and honeybee taxonomy finally ended up in a confusing multitude of inadequately described units. Effective methods of morphometric-statistical analysis of honeybee populations, centered on limited areas, have been developed during the last decades. Only the numerical characterization of the populations, together with the description of behavior, shows the true geographic variability and will end current generalizations and convenient stereotypes.

This book attempts to achieve a synopsis of all available morphometric, behavioral, and ecological data of the known geographic variants of *Apis mellifera*. For more than 25 years samples were systematically collected and analyzed, following a procedure which was elaborated in preliminary tests. Finally, data from more than 1200 samples, workers and drones, with a total of about one million single data, were available for statistical analysis. Some of the results were not published previously. The data and the collection of samples are stored at the Institut für Bienenkunde in Oberursel (Polytechnische Gesellschaft), University of Frankfurt. Information on behavioral and ecological characteristics was found in beekeeping journals, unpublished reports, and also in general press reports, apart from scientific publications, due to the peculiar situation of the honeybee in biology, apiculture, and public interest.

To understand fully the biogeography of *Apis mellifera,* it was necessary to study the evolution and, comparatively, the morphology and biology of the other *Apis* species, resulting in short monographies of *Apis florea, A. dorsata,* and *A. cerana.* In final analysis, the biographical synopsis of the Western Honeybee necessarily ended up as a natural history of the genus *Apis.* The information provided may help to integrate results of special investigations into the general pattern of honeybee biology and taxonomy and to prevent biologically unfounded developments in apiculture. The geographic variation in honeybees furnishes interesting examples of climatic adaptations within one and the same species, patterns of distribution and isolation and various levels of speciation.

The writer of a publication which relies on the contribution of many co-workers feels himself rather as the focus for the efforts of others than as a true author. No overview of the geographical variability of honeybees – although still incomplete – could have been achieved without the kind help of the many colleagues who collected the bee samples, sometimes under difficult conditions. Since it is impossible to list them individually, I want to name only one of them, as representative of all others, to whom I owe the greatest individual contribution: Brother Adam of St. Mary's Abbey, Buckfastleigh, UK, whose collection from his journeys round the Mediterranean furnished the basis for the data bank. During the long years which preceded published results, we relied on the patient assistance with finances and organization by the Deutsche Forschungsgemeinschaft (Bonn) and the Polytechnische Gesellschaft (Frankfurt), the generous sponsor of the Institute in Oberursel.

For contributions to this study some special acknowledgements are called for: to Agnes Mohr, the indispensable aide who patiently made most of the measurements and designs; to Dorothea Kauhausen, who took care of the statistical analysis; to Christl Rau for many of the photographs; to Aasne Aarhuis for several artistic designs. Further I wish to thank numerous authors, cited at the respective places, for the kind permission to use their illustrations. I am grateful to Howell Daly, Martin Lindauer and Tom Seeley for reading specific chapters and making valuable suggestions. This manuscript was written in close cooperation with Hedi Langfeldt, who worked as unofficial editor and sometimes even as co-author.

Lunz am See/Oberursel, Autumn 1987   FRIEDRICH RUTTNER

Contents

Part I   Honeybees of the World

CHAPTER 1

The Genus Apis

1.1 Introduction

The subfamily Apinae with only one tribe, Apini, comprises not more than four (or perhaps five or six) species (Fig. 1.1). The spectacular differences in size of individuals and nest architecture between the species caused taxonomists to subdivide the tribe in several genera or subgenera (Megapis, Micrapis, Sigmatapis) each with one (or two closely related) species (see Chap. 5). Since any taxon, except the species, is a question of deliberate agreement intended for improvement in practical application (Mayr 1968), it seems preferable to use the least complicated classification, that is to group the few species as one single genus.

1.2 Differentiation of Apinae

The subfamily Apinae is discriminated from the other Apidae by only a few minor morphological criteria, but by essential behavioral characteristics listed below:

Fig. 1.1 The four honeybee species: *Apis mellifera, A. cerana, A. dorsata, A. florea* (from left to right). (Photo Institut für Bienenkunde, Oberursel)

A. Morphology

   1. Membranous endophallus.
   2. Stretching of the pattern of wing venation (elongation of marginal and sub-marginal cells of fore wing).

B. Behavior

   3. Vertical combs with hexagonal cells constructed bilaterally and exclusively from self-produced wax. No special nest cover.
   4. Tendency for multifunctional, repeatedly used cells.
   5. Clustering behavior as an essential part of biology and evolution ("contact type" vs. "distance type" of Meliponinae. Sakagami 1971).
   6. Progressive feeding of larvae.
   7. Communication and recruiting by "dance language".
   8. Nest cooling by evaporating water collected in the field.

The characteristics listed above are different from those of Meliponinae, which on the other hand possess several common traits with Apinae: lack of spur on hind tibia (a reminder of the ancestral type is preserved in a pocket of the pupal cuticle, the spur sheath – Chap. 3, Fig. 2.3), perennial colonies with large population, morphologically different queen and worker castes and intra-colony coordination by elaborate chemical and behavioral communication. Winston and Michener (1977) published convincing evidence that eusociality evolved independently in both subfamilies, in spite of some analogies (Chap. 3). Each of them achieved similar high levels of social organization within the Apidae. Rather primitive types are known in Meliponinae, while each of the *Apis* species is equipped with highly derived characters of its own and lower-level types no longer exist. *One Apis*-specific characteristic probably provided the prerequisite for adaptation to zones of temperate climate: the behavior of clustering as the main factor of protecting the nest and regulating its microclimate. The cluster is a permanent feature of open-air nesting species ("protective curtain"), evidently competing well with the method of constructing nest covers in sympatric Meliponinae. Clustering occurs periodically in cavity nesting species of *Apis* (winter cluster, reproductive and migratory swarms).

   The second prerequisite for increasing independence from variable environmental conditions is the unique system of communication. The ecological efficiency of the "information-center strategy", which is basically the same in all *Apis* species, was demonstrated by Seeley (1985). The foraging force of a *mellifera* colony daily adapts anew to the changing availability of food resources in an area up to 6 km (in special cases even to 10 km; v. Frisch 1967) around the nest, thus securing an optimum exploitation of the supply. For populations living in marginal zones of existence (p. 172) this may frequently be the crucial point. While studying honeybees in all biotopes, especially in those of the most extreme kind, the full biological significance of the precise intra-colony communication about available food sources becomes evident. Sometimes a few days or even hours fully exploited for food collection suffice for the survival of a colony.

1.3 Diversification Within the Genus

The diversification within the genus *Apis* concerns all categories of characters (Fig. 1.1, Table 1.1): morphology (size, hair, wing venation, copulatory organ), behavior and distribution. The most conspicuous differences found within the genus concern size of workers (Fig. 1.1, Table 1.2, 1.3) and nesting behavior. Fore wing length taken as a measure of size, a factor of 2.4, gives the difference between the smallest *(A. florea* = 6.06 mm) and the largest bee *(A. d. laboriosa* = 14.50 mm). Thoracic mass was found to be 7.6 mg in *A. florea* and 45.5 mg in *A. dorsata* (= 5.98 x; Dyer and Seeley 1987). Three groups of non-overlapping size are found: (a) *A. florea*, (b) *A. cerana-mellifera*, (c) *A. dorsata* (Fig. 4.2).

The *size of the workers* is evidently adapted to ecological requirements: the Dwarf Honeybee lives mainly in dense bushes, the Giant Honeybee on tall trees or cliffs, its size being an important factor in specific defense behavior (Seeley et al. 1982). The medium-sized bees of the *cerana-mellifera* group are apparently the result of a compromise between the foraging range needed for provisioning, size of nesting sites, and regulation of microclimate. The correlation of worker size with geographic latitude and altitude found at least in *A. mellifera* and *A. cerana* (p. 53, 152) can be interpreted in the same sense. Within the species *mellifera* a number of subspecies are discriminated by size alone (Fig. 4.3). Variation within *A. cerana* is probably of the same range, taking into consideration the fact that only a very incomplete survey was done on this species (Table 9.6). Size alone, therefore, is an insufficient criterion in bee taxonomy: non-overlapping groups are also frequently found within species.

The interpretation of size of individual worker bees as a partly functional adaptation is corroborated by the observation that differences between sexuals of the species are distinctly smaller than differences between workers. Species with small workers have relatively large sexuals *(A. florea*, Figs. 7.2, 7.3) and vice versa *(A. dorsata*, Fig. 8.6, Table 1.2). There is a fivefold range in worker body mass, but less than a twofold range in the size of queens and drones (Table 1.2; Dyer and Seeley 1987).

Nesting Behavior. The classical description of the various nest types provides the basis for the traditional classification into four species:

A. Single-comb open-air nest   B. Multiple-comb cavity nest
A. florea   *A. cerana*
A. dorsata   *A. mellifera*
(A. laboriosa)

This rough division of the genus based on nesting behavior makes sense also because of the ecological implications: group A is restricted to the tropics and subtropics, while the two species of group B were able to colonize the cool temperate zone without losing the ability to compete in tropical climate. Recent investigations by Dyer and Seeley (1987) indicate that essential physiological differences exist between the two groups: the cavity-nesting species *A. cerana* and *A. mellifera* fly with higher speed compared to *A. florea* and *A. dorsata*, they have a higher thoracic temperature (= higher heat production in consequence of a greater

Table 1.1 Species-specific characters of *Apis*

Character	Florea	Dorsata	(Laboriosa?)	Cerana	Mellifera
Fore Wing length (mm)	6.0–6.9	12.5–13.5	14.2–14.8	7.27.–9.02	7.64–9.70
Tomenta	Tergite 3–6	3–6	3–6	3–6	3–5
Hind wing: extension of radial vein	Variable	Present	Present	Present	Missing
Melittin: sequence of amino acids (deviation of *mellifera*-type)	5 Amino acids changed	3 Amino acids changed	?	0	0
Drone					
Endophallus	1 Pair of cornua; bulb a thin tube	4 Pairs of very long thin cornua; short bulb	?	1 Pair of cornua, rudiments 3 others no chitin. plates	1 Pair of cornua bulb with chitin. plates
Basitarsus 3	deep incision with plumose hair+spines	Thick pad of sturdy branched hair	?	Thin pad of plumose hair	As cerana
Behavior					
Capping of drone cells	Solid	Solid	?	Perforated	Solid
Nest	Single comb encircling twig to form a "dance floor", fixed with cell bases	Single big comb fixed at bottom side of branch or rock, fixed with midrib	as *dorsata*	several combs in cavity, fixed with midrib	As *cerana*
Communication	Sun-oriented dance on platform open to the sky	S.-o. dance on vertical comb open to the sky	?	S.-o. dance on vertical comb in cavity	As *cerana*
Distribution	Sympatric	Sympatric	?	Sympatric	Allopatric

Table 1.2 Measurements of size and hair length (mm), data on proportions, body mass and number of wing hooks of four honeybee species. (Data on body mass from Dyer and Seeley 1987; data on fore wing length and width see Table 4.3) "n" equals number of individual bees (not samples)

Species	n	Hair length	Tongue	3. Stern. longit.	Wax plate transvers.	Sternite 6 long.	Sternite 6 trans	St6 Ind	Hind leg	Cub. Ind	Hooks hind w.	Body mass
A.florea (Oman)	20	0.118	3.31	1.83	1.48	1.45	1.88	77.6	5.20	2.82	12.9	22.6
A.cerana (Pakistan)	140	0.287	5.40	2.48	2.14	2.31	2.80	82.6	7.58	3.98	16.86	43.8
A.mellifera (ssp.mellifera)	195	0.418	6.11	2.84	2.43	2.62	3.35	78.5	8.09	1.82	21.7	77.2
A.dorsata (Pakistan)	10	0.242	6.45	4.08	2.55	3.12	3.05	102.1	10.27	7.25	24.30	118.1

Table 1.3 Size comparison for the sexuals and workers within and between four honeybee species, in mm. (Partly with data of Seeley et al. 1982)

	Workers Head width	Wing length	Queens Head width	Wing length	Drones Head width	Wing length	Ratios Head width ☿ : ♀ : ♂	Wing length ☿ : ♀ : ♂
A. florea (Thailand)	2.60 ± 0.03	6.26 ± 0.10	3.19	8.49	3.73 ± 0.04	9.23 ± 0.13	1: 1.23: 1.43	1 : 1.36: 1.47
A. cerana (Tailand)	3.38 ± 0.06	7.54 ± 0.14	3.65	9.41	3.60 ± 0.04	9.01 ± 0.11	1: 1.08: 1.07	1 : 1.25: 1.19
A. mellifera (*mellifera*)	3.77 ± 0.04	9.32 ± 0.16	3.75	9.92	4.52 ± 0.07	12.24 ± 0.51	1: 0.99: 1.20	1 : 1.06: 1.31
A. dorsata (Thailand)	4.71 ± 0.09	12.34 ± 0.34	4.81	12.78	4.60 ± 0.07	13.35 ± 0.26	1: 1.02: 0.98	1 : 1.04: 1.08

power output) and a heavier "wing loading". This implies a higher metabolism rate of *A. cerana* and *A. mellifera*, evidently correlated with a shorter life span of the worker bees of these species (see p. 86, 109). According to Dyer and Seeley, the two cavity-nesting species show a high-tempo existence with a high turnover in the colony, in contrast to a low-tempo, long-lived existence of *A. florea* and *A. dorsata*. It should not be overlooked, however, that the phylogenetic relations between these species are probably far more complex than is suggested by this simple splitting in two. Comb architecture is the same in *A. dorsata* and *A. cerana/mellifera* while it is basically different in *A. florea* (Chap. 7; Ruttner et al. 1985a).

1.4 Classification

Historical development and the present state of taxonomic classification of honeybees are described in Chapter 4. Starting point of any argumentation is the highly polytypic but well-studied species *A. mellifera* (Chaps. 10–14). The broad geographic variation of this species results in subspecies strongly diverging in various characters including size. By numerous hybridization experiments it has been proven that even the most distant subspecies belong to one and the same species. The subunits (= geographic races) of *A. mellifera* occur exclusively allopatrically and produce fully fertile hybrids if crossed with each other. Since the differences between the races of the other three species are of the same kind and dimension, the results obained with *A. mellifera* can be extrapolated to these species.

This well-documented intraspecies variability of *A. mellifera* has to be regarded as exemplary for selecting species-specific characters in the genus. Important quantitative differences can be secondary attributes, but they are not evidence for species recognition. The four traditional *Apis* species, *florea, dorsata, cerana,* and *mellifera,* are characterized by certain *qualitative* body structures, especially in the male genitalia, which are absolutely specific for each of the four species and never found in any one of the others. Differences of this kind exist even between the two young, not fully mature species, *cerana* and *mellifera*. Of course, important *quantitative* differences are found between species as well (Table 1.1), but they may sometimes strongly overlap.

1.5 Apis laboriosa?

Sakagami et al. (1980) published a morphometric analysis of a "super *dorsata*" bee from altitudes between 1500 and 4000 m in Nepal, using more than 70 selected characters. This bee was first named "Megapis laboriosa" by Cockerell in 1906. The proposition to recognize this taxon as a fifth *Apis* species is based on the following arguments: (1) Important quantitative differences supposed to be exceeding those found within one of the known *Apis* species (Fig. 1.2, Table 1.1). (2) Presumably sympatric occurrence. (3) Ecological divergence (Roubik et al. 1985).

These arguments have to be seen together with the well-documented and analyzed situation in *A. mellifera*: the subunits of this species differ from each other in size, hair, and wing venation as much as *laboriosa* does from *dorsata* s. str. (Table 1.1, Chap. 4, p. 37). The consequence of colonizing a new ecological niche is a considerable morphological diversification, also at the subspecies level. Increasing knowledge about geographic variation in *A. mellifera* reveals quite a few examples of intraspecies ecological and morphological diversification and isolation (e. g., *A. m. litorea* – *A. m. monticola*; "Africanized" bees).

Differences regarded as species-specific in the *dorsata* group are found on the same order of magnitude at the subspecies level in *A. mellifera* (Fig. 4.2). No single major qualitative characteristic has been presented for *laboriosa*. Moreover, Delfinado-Baker et al. (1985) showed that *laboriosa* and *dorsata* share the same parasites; this is not observed in other *Apis* species. On the other hand, two new *Braula* species of the genus Megabraula were recently detected in *laboriosa* nest which

Fig. 1.2 *Apis laboriosa* (top center) with two subspecies of *A. dorsata* (*breviligula* and *binghami*). Bottom: *A. cerana* (left) and *A. mellifera*. (Photo F. Sh. Sakagami et al. 1980)

may or may not occur in *dorsata* nests (Grimaldi and Underwood 1986). Although very little is known about the biology, *laboriosa* certainly exists under extreme ecological conditions, with the consequence of typical morphological changes (p. 118). The taxonomic rank is determined not by the morphometric distance but by the genetic isolation and historical age of the type. The problem "A. laboriosa" will be instantly solved as soon as the male copulatory organ is described in detail (according to a personal communication by B. Underwood current investigations show that the endophallus of *laboriosa* is identical with this bizarre structure in *dorsata*) and the allopatric distribution established. The extremely complicated structure of the *dorsata* endophallus is expected to be soon modified in genetic isolation.

More important than this question of classification are the problems of physiology and ecology of this most exceptional yet little known honeybee. All data available indicate a further step in evolution to the limits of adaptability of an open-air nesting species: existence in a zone with temperate climate, surviving temperatures below freezing point at least for a limited time (Roubik et al. 1985). The direction of evolution points to a *Bombus* type rather than to a multi-comb *Apis*: large body size, long hair, dark color.

In conclusion, sufficient data are available now to recognize four species within the genus *Apis*. A possible fifth species, closely related to *A. dorsata* and living in an extreme biotope, is not yet fully confirmed.

1.6 Behavioral Isolation of Species

Three *Apis* species, highly different in their morphology, exist sympatrically in South Asia. Although all drones respond to the same sexual attractant, the three species are genetically isolated by different mating times. In Sri Lanka drones of *A. florea* show maximum flight activity for 2 h during the early afternoon (12.30–14.30). *Cerana* drones fly between 16.30 and 17.30. The flight period of

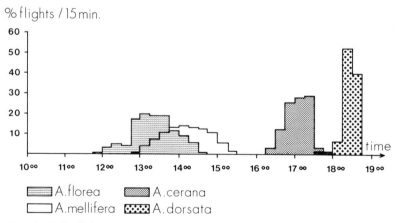

Fig. 1.3 Daily flight activity of drones of *A. florea, A. cerana, and A. dorsata* in Sri Lanka, and of *A. mellifera* in Louisiana, USA. (Koeniger and Wijayagunasekera 1976)

A.dorsata lasts less than 1 h shortly before dusk (Fig.1.3; Koeniger and Wijaya-gunasekera 1976). In Europe, flight periods of imported *cerana* drones and local *mellifera* drones greatly overlap, and hybrid matings are attempted (p.148).

1.7 Food Competition

Koeniger and Vorwohl (1979) observed the interactions among four species of Apidae *(Trigona, A.florea, A.cerana* and *A.dorsata)* at an artificial feeding dish. Numerous interactions among the species were noted, small individuals generally attacking larger ones. *A.dorsata,* however, was attacked only by *A.cerana,* never by the other two species. At times, only one species finally remained while the others stayed away. It was not predictable which would be the "victor". Smaller species have less territory, which is defended by more aggressive behavior. *Dorsata* honeys contain the lowest number of different pollen, probably from highly rewarding plants farther away. Therefore it seems that the bigger species avoid disastrous competition by shifting to other, more distant sources. Bees of the same *mellifera* colony behave in the same way late in the season when nectar supply is getting poor (Visscher and Seeley 1982). Koeniger and Vorwohl conclude that competition for food is not the limiting factor for the existence of the bee species studied. This is confirmed – at least in average years – by practices of beekeeping when a very high density of honeybee colonies is tolerated without visible effects. Of course, large differences between regions exist. The competition between different groups of honeybees and various species of Apoidea ("natives" = *Xylocopa, Bombus, Halictus* and others) on a crop of *Agave schottii* was studied by Schaffer et al. (1982). When additional honeybee colonies were brought into the area, the number of flower-visiting local honeybees and "natives" diminished at the same time as the amount of nectar measured in the blossoms. When the additional colonies were removed, the number of all groups of "locals" increased again.

1.8 Ecological Adaptations to the Tropics

A thorough comparative study of three *Apis* species was undertaken by Seeley et al (1982) in a tropical rain forest of Thailand, focusing on the impact of predators. A specific defense strategy was established for each species (see Chaps.7-9). *A.florea* proved to be the most vulnerable: while only 10% of the *cerana* nests were destroyed or abandoned per month (on the average), 25% of *florea* nests perished during the same period. Since *florea* continued to exist in this region it must have made up for the losses by absconding, by a higher reproduction rate, and by superior availability of nesting sites.

Predation is probably a heavy selection pressure on all tropical honeybees. In dry regions (Africa) long-lasting droughts and bush fires may be of equal significance. On the other hand, it is mainly climate, particularly the cold season, which creates problems for honeybees in the temperate zone. In the cold temperate zone (State of New York) 90% of colony mortality occurs during winter, presumably mainly due to starvation (Seeley 1978).

As a result, two different behavioral patterns are found in the two climatic zones:

- In the tropics a migratory way of life, high reproduction rate and pronounced defensive tendencies.
- In the temperate zone a strictly stationary life, low reproduction rate, weak to moderate defense behavior and the ability to survive a period of 5 months or more without any flight activity.

It is a peculiar phenomenon that these fundamental features are not completely correlated with the taxonomic classification. Evidently, there are two exclusive tropical species, the open-air-nesting *A. florea* and *A. dorsata*. Yet the other two multi-comb species comprise both tropical migratory and stationary subspecies with low reproductivity. The reason for this astonishing situation is the comparatively recent origin and close relationship of the two honeybee species which colonized the temperate zone. This change, although occurring *within* species, represents a big evolutionary step. This is impressively shown by an unplanned experiment of enormous dimensions, the encounter of tropical African with European honeybees in South America. They behave almost like separate species, but they are isolated not by genetic, but by behavioral factors.

The "tropical" characters of honeybees are described in detail in the special Chapters 7–10. The adaptation to a cold climate concerns only *A. mellifera* (Chap. 10) and (the not yet investigated) *A. cerana*.

The survival of honeybees in temperate climate is, of course, primarily a question of temperature regulation, but that alone is not sufficient. To be able to live through several months or more of permanent cold, a whole set of characters had to evolve. In a warm temperate climate like the Mediterranean and the Near East, the flightless periods generally last only a few weeks at the most. In regions with a humid winter season, nectariferous flowers are available at the beginning and end of the cold season. Only strains of a cold temperate zone show the full scale of wintering adaptations. Therefore, not more than a few races of *A. mellifera* can be regarded as fully adapted: *A. m. mellifera* (West, Central, and NE Europe), *A. m. carnica* (SE Europe) and probably *A. m. anatoliaca* (Br. Adam 1983). Even within these races differences exist between northern and southern populations (South France and NE Europe within *mellifera*; Alpine and Adriatic populations within *carnica*). The special adaptations of these races are discussed in Chaps. 10, 11, 13, 14.

CHAPTER 2

Stingless Bees (Meliponinae)

Bombinae and Meliponinae are the only reasonably close relatives of honeybees. Since Meliponinae have many biological characteristics in common with Apinae (storage of honey, production and use of wax, perennial colonies of sometimes considerable size, division of labor), and since workers of some of the (bigger) species at first glance greatly resemble *Apis* workers (Fig. 2.1), they were even named "stingless honeybees" in contrast to the "true honeybees", *Apis* (Winston and Michener 1977).

To this list of similarities, however, an even longer list of dissimilarities can be added (Table 2.1).

From the divergence in morphology and ethology Winston and Michener (1977) concluded that Apinae are not a specialized offshoot of Meliponinae, but that highly eusocial behavior evolved independently in both groups. Similar or

Fig. 2.1 Queen and workers of *Mellipona favosa.* (Photo N. J. Sommeijer)

Table 2.1 Characters of Meliponinae, different from *Apis*

Morphology	– Reduction of sting apparatus and wing venation (Fig. 2.2) – Mainly small size (down to 2 mm) – Penicillium (= rows of long curved bristles on the apex of the tibia) but no auriculum on the metatarsus – Wax glands at abdominal tergites, etc.
Behavior	– See Table 2.2
Genetics and Phylogeny	– Chromosomal polymorphism – Advanced stage of cladogenesis

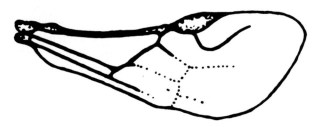

Fig. 2.2 Fore wing of *Trigona (Trigonisca)* sp. with reduced wing venation. (Wille 1979)

corresponding traits are regarded as a consequence of parallel or convergent evolution.

Since among recent Apinae no true primitive types are found , the evolutionary radiation being small or, more precisely, being in its initial stage, a comparison with the strongly diversified Meliponinae could provide conclusions on evolutionary trends in Apinae. This is the main purpose of this short outline of taxonomy, distribution, and biology of Meliponinae.

At present, more than 500 species of Meliponinae have been described (Roubik 1987). However, "every expedition to Brazilian areas where few bee collections have been made reveals several new species" (Kerr and Maule 1964). The number of genera set up varies with the tendencies of the respective author (whether he is a "splitter" or a "lumper").

A. Wille (1979), regarding himself of intermediate position in this respect, recognizes 8 genera and 15 subgenera.

Figure 2.3 shows the degree of radiation within the Meliponinae, presenting a striking difference when compared to the low level of morphological and behavioral differentiation within the Apinae. In contrast, however, the range of ecological adaptation is much wider in the Apinae. The Meliponinae occur in all continents, but they are restricted to the circumequatorial zone (Fig. 2.4). No stingless bees occur in the temperate zones.

The geographic distribution of Meliponinae seems to indicate a center of speciation in South America (Table 2.2).

A. Wille (1979), however, presented a number of arguments which strongly suggest that in spite of this recent distribution of species the center of origin of Meliponinae had been Africa:

Table 2.2 Geographic distribution of Meliponinae species. (Kerr and Maule 1964)

Region	No of species
South America	183
Africa	32
Asia	42
Australia, New Guinea, Solomon Islands	20

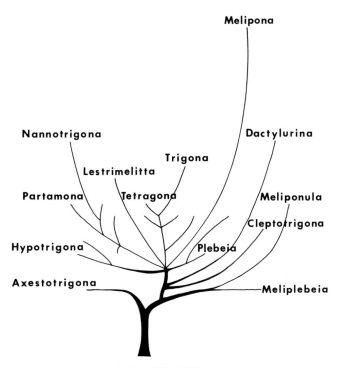

Fig. 2.3 Radiation of the subfamily Meliponinae. (After Wille 1979)

The earliest fossils (Eocene) were found in Europe, which was connected with Africa at this period, while South America was completely isolated. The African species of Meliponinae are primitive. Only there (and not elsewhere) are species with a sting and a flat gonostyle found, both regarded as primitive characters.

A. Wille (1979) concluded that Meliponinae evolution started in Africa, but was impeded by unfavorable climatic conditions during the aridity in the late Tertiary, which possibly even reduced the number of species. In South America, on the other hand, Meliponinae speciated profusely in a highly varied, very favorable environment. Population density there can be extremely high. Roubik (1983), in cut trees of the cleared forest in Panama, counted 30 nests belonging to 14 Meliponinae species ("and there were undoubtedly more").

The area of distribution of Meliponinae coincides almost exactly with the tropics (Fig. 2.4). In spite of their evident genetic plasticity and of drastic climatic

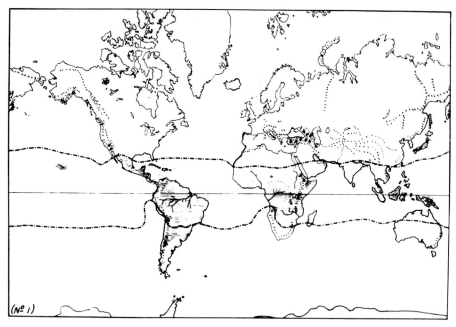

Fig. 2.4 World distribution of Meliponinae. (Kerr and Maule 1964)

changes, the Meliponinae were not able to evolve species adapted to zones with temperate climate. This is a remarkable phenomenon, taking into consideration that the level of sociality in many respects equals that of *Apis*, or is even superior (Table 2.3).

This list of single factors of social behavior important for fitness gives no clue for the reason of restriction in ecological radiation in Meliponinae. Surprisingly, the character which potentially could be of very high adaptive value in cool climates, a thermo- and humidity-insulating cover, is found in Meliponinae, but not in *Apis*.

There is, however, one ecological-behavioral difference, which might be the crucial point: thermoregulation within the colony. In honeybees, especially in *A. mellifera*, thermoregulation was extensively studied (see Heinrich 1985), and it was shown that its temperature homeostasis compares fairly well with that of warm-blooded vertebrates. Meliponinae, though only partially investigated, are far from this level of regulatory perfection. Apparently only 13 of the more than 400 stingless bee species have the ability to regulate nest temperature within certain limits, and these are species with large colonies in voluminous nests (Roubik pers. commun.). The involucrum surrounding the brood helps to maintain nest temperature above ambient temperature; however, no temperature homeostasis was found so far. There is no evidence that stingless bees carry water to the nest for evaporative cooling (Roubik 1987). Various species endure low temperatures, but only for a short period, because they have to clean the nest within short intervals and they do not cluster; therefore they would not survive in cold climates.

Table 2.3 Characters of sociality and fitness of Meliponinae compared with *Apis* (Sakagami 1971, A. Wille 1983)

Character		In Meliponinae
Number of workers	–	About equal in M. species with large colonies
Differentiaton of castes	–	Equal
Division of labor	–	Present, though less developed in the few cases investigated (Sommeijer 1983): differentiation between young bees (within the nest) and old field bees
Brood care	–	Mass provision
Reproductive swarms	–	"Progressive" type of starting a new colony without swarm cluster beginning with selection of a new nesting site; during gradual transfer of part of the bees with a young queen (lasting up to several months) transport of nesting material. Swarming distance only up to 300 m – however, distribution round the globe
Interindividual relations	–	"Distance type" (no cluster formation; Sakagami 1971; see p. 4, 33)
Communication about food sources	–	Present, but not as sophisticated as in *Apis* (alerting, guiding, odor marking)
Storage of food	–	Important quantities observed even in hot, humid climate with indication of antifermentic agent, preserving honey with high water content (Roubik 1983)
Flight range	–	Established up to 2000 m (Roubik 1983, A. Wille 1983), that is farther than in tropical *Apis* (Lindauer 1956)
Nest structure	–	Definitely better adapted in highly specialized species, with an insulating multi-layered closed cover (e.g., *Trigona nigerrima*, (Fig. 2.6). There are observations on survival of pupae in nests flooded for several months (Roubik 1983)
Temperature regulation	–	Present, but no true homeothermy (Fig. 2.5; Roubik 1987)

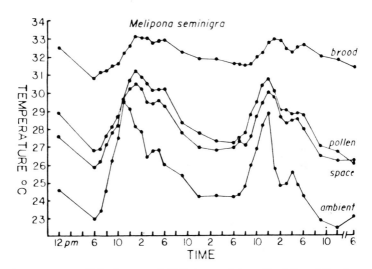

Fig. 2.5 Regulation of temperature within the nest of *Melipona seminigra*. (Roubik and Peralta 1983)

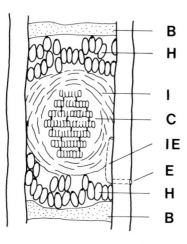

B
H
I
C
IE
E
H
B

Fig. 2.6 Nest of *Trigona nigerrima* in a hollow tree trunk (Kerr et al. 1967). *B* Batumen, *C* Brood cells, *E* Entrance, *IE* Entrance tube, *I* Multilayered involucrum, *H* Honey pots

Colonies of *Trigona* were kept for several winters in Germany, maintained at room temperature with free nest entrance to the outside and permanent care (Engels pers. commun.).

Meliponinae, despite the ability to build protecting nest covers, but with only vestigial attempts at temperature control, are restricted to the region of the tropics. A*pis mellifera*, on the other hand, substituting the lack of nest covers of its own by only poorly adapted natural cavities, but with perfect thermohomeostasis, was able to colonize vast regions throughout warm temperate to cold temperate zones. European honeybees spread quickly over the whole continent of North America when imported by farmers. They found an empty paradise in which to multiply by the millions; but this paradise has been closed to Meliponinae living in Central America for many millions of years, simply for lack of ecological fitness.

This enormous extension of biosphere of two species of a single genus, within a short geological period, evidently due to improved thermo-regulation, shows the fundamental significance of this physiological-behavioral accomplishment. The increase in independence from environment by the achievement of thermoregulation is regarded as one of the major steps in vertebrate evolution. The same effect is seen in social insects. To make this decisive step has evidently been as difficult and tedious in these animals as it was in vertebrates.

Only some members of the stingless bee family store honey in quantities that can be harvested by man: in South and Central America several species of Meliponinae are exploited by Indians. In several regions techniques of beekeeping with Stingless bees were developed, using calabash (Paressi Indians in Mato Grosso), hollow logs (Menimehé, Rio Gapura) and pottery (Colombia and Venezuela) as beehives (Nordenskjoeld 1934). In Guatemala, cylindrical wicker hives are fixed to the house, just as can be seen with colonies of *A. mellifera* and *A. cerana* in different parts of Asia. In East Africa, colonies of Stingless bees are placed close to the house and the honey is harvested by the owner (pers. observ.). The most sophisticated "meliponiculture" was developed (and is practiced even today) by the Mayas in Yucatan, with a specific ritual dating from precolumbian times (N. and E. Weaver 1981, De Landa 1566; Fig. 2.7). The tradition of meliponiculture

Fig. 2.7 Picture of meliponid from the Maya period:
emblem of the 27th International Beekeeping Congress
1981 in Acapulco

is well documented by the report that Cortez found thousands of hives used by
Indians when he first landed on the island of Cozumel in 1519 (Nordenskoeld
1934). There was an "abundance of honey and wax" (De Landa 1566). In Amazo-
nia (Manaos), *M. seminigra* is reported to produce more honey than *A. m. ligustica*
(Renner 1982).

CHAPTER 3

Evolution

3.1 The Evolution of the Main Subfamilies of Apidae

The study of honeybees provokes questions as to their origin. The subfamily Apinae belongs to a family of bees living in social communities, the Apidae, characterized by special organs for pollen collection (pollen comb, pollen basket). No
intermediates, less perfect in their social structure, link the four existing, highly
eusocial *Apis* species to other subfamilies. This is in contrast to the full scale of
intermediates of the many stingless bee species. Showing a superficial morphological similarity, the Meliponinae were for a long time taken as the closest relatives
and ancestors of Apinae (Michener 1944, 1974). A later analysis resulted in a different phylogenetic dendrogram (Winston and Michener 1977). It seems useful to
discuss this problem by considering various aspects.

3.1.1 Fossils

In the rich insect fauna of the Baltic amber (upper Eocene in the early Cenozoic,
40–50 m. y.) many Hymenoptera were found. No *Apis* or *Bombus* species were
among them, but different species of the recent meliponine genus, *Trigona* (Kel-
ner-Pilault 1969; Zeuner et al.1976). Of special interest is a peculiar bee of the
Baltic amber, first described as *Apis meliponoides* by Buttel-Reepen (1906), and
later classified as the separate genus *Electrapis* (Cockerell 1908), although it is very
different from the present genus *Apis*. It shows a number of variable, intermediate
characteristics of all the three subfamilies of Apidae, e. g., indication of a developing auricle on hind basitarsus (Fig. 3.1), one tibia spur as opposed to none in *Apis*
and two in Bombini, shape of body or wing venation. No wonder that it is a permanent source of annoyance for taxonomists; their classification varies according
to the specimen and the characters examined (see Kelner-Pillault 1969; Kerr and

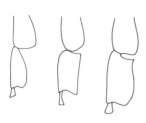

Fig. 3.1 Basitarsus of hind leg of *Melipona, Electrapis* and *Apis*,
from left to right. (Buttel-Reepen 1906)

da Cunha 1976; A. Wille 1977). The situation is best described by Michener (see Winston and Michener 1977) when mentioning a specimen of *Electrapis (Roussyana) proava* (Menge): "... it would seem to be an apine in spite of being superficially *Trigona*-like. Its wing venation, however, is *Bombus*-like ...". Zeuner et al. tried to overcome the dilemma concerning this polymorphic, ambiguous genus by creating three subgenera, one covering types similar to each one of the three subfamilies of Apidae:

Electrapis (Electrapis), apine-like,
E. (Protobombus), bombine-like,
E. (Roussyana), meliponine-like.

In conclusion, "*Electrapis*" seems to have been a group of species in an ancestral position in respect to all recent Apidae. At the same time, however, typical Meliponinae existed already. Therefore, they have to be considered as the most ancient subfamily.

3.1.2 Morphology

In spite of a clear characterization of the subfamilies, the deduction of a plausible phylogenesis is not easily achieved at first sight. The reason for this is a seemingly irregular distribution of original apoid and derived characteristics among subfami-

Table 3.1 Distribution of characters among subfamilies of Apidae (Winston and Michener 1977). + = present, − = absent

Taxa	No. of derived charact.	Derived character	No. of primit. charact.	Primitive (parental) character
1. Apinae	4	1. Endophallus 2. Temperature homeostasis 3. Progressive feeding 4. Same cells for brood and provisions	0	–
2. Apinae and Bombinae	5	1. Basitarsus: auriculum + 2. Fore-wing: stigma − 3. Wax glands ventral 4. Maxilla : stipital process 5. Heterolateral transfer of pollen	2	Wing venation Sting
3. Apinae and Meliponinae	4	1. Hind tibia: spur − 2. Slender form 3. Large perennial colonies 4. Swarming	1	Jugal lobe

Common characteristics of Apidae: 1. Corbicula (pollen basket) 2. Rastellum (pollen comb) 3. Pollen loading from distal end of tibia.
It should be noted, however, that contrary to this clear-cut tabulation "mixed" types occur. Rudimentary tibia spur in the pupal cuticle of *Apis*; perennial colonies and adult-larvae contact in *Bombus*; additional dorsal wax glands in *Bombus*.

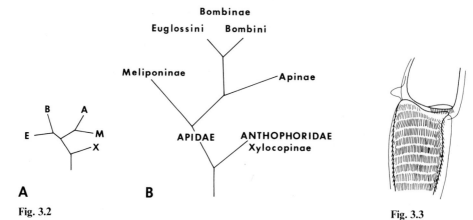

Fig. 3.2

A

B

Fig. 3.3

Fig. 3.2 Cladistic structure of *Apidae*. A (Michener 1941, 1974). B (Winston and Michener 1979)

Fig. 3.3 Rudimentary tibia spur at the hind leg of *A. mellifera* workers: empty pocket in the pupal skin. (Leuenberger-Morgenthaler 1954)

lies and tribes. Listing the characteristics given by Winston and Michener (1977) and adding a few others, a rather even distribution of derived characters among the following three groups is found (Table 3.1): *Apis* alone (males with endophallus, identical cells for brood and provisions); *Apis* + Bombini (basitarsus with auricle, stigma of fore-wing missing); *Apis* + Meliponinae (tibia spur missing, large perennial colonies). This mosaic of characters opens the way for diverging interpretations: the conventional cladistic dendrogram (Fig. 3.2; Michener 1974) was modified with convincing arguments by shifting Apinae from the neighborhood of Meliponinae to Bombinae (Winston and Michener 1977). Whatever the similarity, the main fact remains: the close relationship of all subfamilies of Apidae, probably best represented by the idea of a descent from a common gene pool in the Eocene, with Meliponinae being a little older. Remnants of ancestral charcteristics may emerge in the most advanced species. The "lost" tibia spur of Apidae is found as an empty pocket of the pupal skin in *A. mellifera* (Fig. 3.3).

3.1.3 Principal Period and Extent of Radiation

Each subfamily radiated at a different time: early Tertiary in Meliponinae (Wille 1980), late Tertiary in Bombinae (Williams 1985), end of Tertiary-Pleistocene in Apinae, as will be demonstrated later. This difference in time of evolution is reflected by differences in rank and number of the taxa: Meliponinae – 18 genera, 500 species; Bombini – 3 genera, about 290 species; Apini – 1 genus, 4 (or 5) species (with many subspecies).

3.1.4 Ecology

Each of the three subfamilies of Apidae is well characterized by its general structure, adapted in a special way to different environments: the predominantly small Meliponinae with more or less large, perennial colonies in sophisticated nests, strictly pantropical; the large, sturdy pubescent Bombini with relatively small seasonal colonies in cool (even arctic and alpine) climate; the generally large, slender Apini, which succeeded in attaining remarkable independence from environmental conditions by a very high level of social organization (homeothermy).

3.2 Evolution Within the Honeybees (Apinae)

3.2.1 General Morphology, Biology, and Cytogenetics

Earlier attempts to deduce the phylogenetic relationship of the four recent *Apis* species can be listed as follows:

1. Nest type. Open-air nesting ("primitive") and cavity nesting (Buttel-Reepen 1906).
2. Karyotype. Open-air nesting species were described as haploid (n = 8; Deodikar and Thakar 1966), cavity nesting species as diploid (n = 16).
3. Level of Caste Differentiation. Difference in size between workers and sexuals (as well as between their brood cells) is greater in *A. mellifera* than in *A. dorsata*, indicating a higher level of social organization in the former (Buttel-Reepen 1903). The same is the case with the yolk protein content of the hemolymph of workers; a low content indicates a high level of sociality (Engels 1973).
4. Level of Dance Communication. According to the faculty of transposing the waggle dance direction from the horizontal to the vertical plane and memorizing the position of the sun, a clear sequence of *A. florea - dorsata - cerana + mellifera* was established (Lindauer 1956).

However, the final conclusions derived from these data are partly inconsistent and contradictory:

1a. Open-air nesting is not necessarily "primitive", it can as well be interpreted as a derived adaptation to tropical conditions of primarily cavity nesting bees (Koeniger 1976b).
2a. The karyotype of all four honeybee species was found to be uniformly n = 16 (Fahrenhorst 1976a,b)
3a. Differences in size between castes are even greater in the "primitve" *A. florea* than in *A. mellifera* (Table 1.3). *A. dorsata*, the species with the largest workers, has sexuals which are smallest in relative size. These data suggest that an optimal body size exists in the genus *Apis*, fairly closely retained in the course of evolution in the sexuals which are centered on high reproductivity in favor of the social community. The worker caste, however, had to adapt itself to its spe-

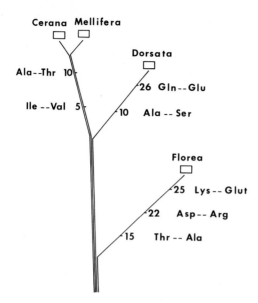

Fig. 3.4 Phylogenetic relationship within the genus *Apis*, derived from differences in the amino acid sequence of melittin. Figures give the position of amino acids. (Kreil 1975)

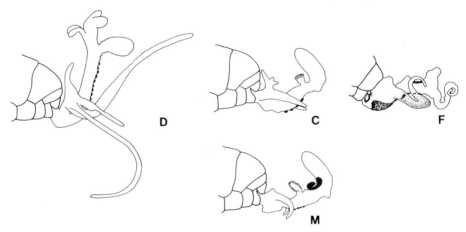

Fig. 3.5 Everted endophallus of *Apis dorsata* (*D*), *A. cerana* (*C*), *A. mellifera* (*M*) and *A. florea* (*F*)

cial biotope, especially nesting site and forage (Seeley et al. 1982). This double trend, acting on one and the same genome, must involve different types of reaction during caste determination – workers diminish or increase in size during evolution, while the sexuals remain to a large extent unchanged.

4a. There is agreement that no transposition of waggle dance direction to the vertical plane occurs in *A. florea* (Gould et al. 1985). However, noncelestial marks visible from the nest with memorized position to the sun are used for dance orientation. "This system implies more complex information processing than in the communication dances of the other species of *Apis*" (Dyer 1985a).

Only three general conclusions result from all these observations:

A) *A. florea* is an isolated, highly specialized type within the subfamily. Structure of the endophallus, architecture of the comb, dance communication, geotactic response etc. are all quite peculiar (Chap. 7). The remote position of *A. florea* was additionally demonstrated by Kreil (1973, 1975) while analyzing the sequence of the 26 amino acids composing melittin, the main protein of bee venom. Differences were found in several positions of the chain (Fig. 3.4): Five amino acids differ between *A. florea* and *A. dorsata*, as well as the *cerana-mellifera* group. Only three amino acids were substituted comparing *dorsata* and *cerana-mellifera*, while the melittin of *cerana* and *mellifera* is completely identical.

B) A sequence *dorsata-cerana-mellifera* can be demonstrated by different characters, most clearly by the morphology of the endophallus (Fig. 3.5). This unique, bizarre organ shows homologous structures in all four species: a membranous tube with the ejaculatory duct inside, various hair plaques and several protuberances of the tube (Fig. 9.11, p. 129). The most conspicuous of these are two long, downward bent "bursal cornua". *A. dorsata* additionally has three pairs of thin dorsal cornua (Fig. 8.7); they are clearly visible as a vestigial structure in *A. cerana* (Fig. 9.11), and even more reduced in *A. mellifera*. Surprisingly, the most specialized organ is found in *A. dorsata,* the most "primitive" species of the sequence. *A. florea* takes a completely separate position. Therefore, no direction of evolution can be determined from these observations.

C) *A. cerana* and *A. mellifera* are very similar species of evidently recent origin, since no pre-mating barrier exists between the two (Chap. 9).

Fig. 3.6 *Synapis henshawi.* Photo R. Snelling

No clear conclusions can be derived from these observations about the evolution of the *Apis* species. The reason is that even the "primitive" species show a rich, elaborate inventory of highly spezialized morphological and behavioral characters and not a "recent adaptive divergence" (Seeley et al. 1982). Therefore, the question of the evolution of honeybees will be investigated from all available data on paleontology, historic and morphometric taxonomy and ecology which were considered essential in the study of the general evolution of Apidae.

3.2.2 The Fossil Record

Well preserved specimens of a true *Apis* type come from the lower Miocene (Rhenish brown coal) and were found in Rott near Bonn, (22–25 m.y. old). They were named *Synapis henshawi* by Cockerell (1907), later classified as subgenus (one species with three subspecies) of *Apis* by Zeuner et al. (1976). From the upper Miocene (Sarmatian, 12 m.y. ago) are the fossil bees of the Randecker Maar and of Böttingen (SW Germany), both classified as *Apis armbrusteri* with several subspecies (Zeuner et al. 1976; Fig. 3.6). Fossil honeybees were found in East African copal and, as they do not differ from recent *A. mellifera*, were assumed to be of Pleistocene origin (Zeuner et al. 1976). In fact, however, they are of very recent origin (Burleigh and Whalley (1983).

A clear trend in changes of wing venation from *Electrapis* to *Apis mellifera* (Fig. 3.7) is observed: (1) Elongation of venation in the length axis of wing; (2) Migration of the third cubital vein along the radialis in the direction of the wing basis. In consequence, the venation pattern becomes definitely more slender.

The honeybee wing venation proved to be a rich source for genetic and taxo-

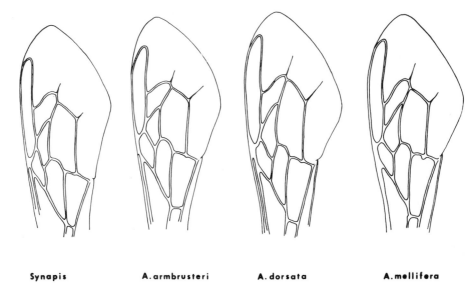

Synapis　　　　　**A. armbrusteri**　　　　　**A. dorsata**　　　　　　**A. mellifera**

Fig. 3.7 Forewing of *Synapis* sp., *A. armbrusteri*, *A. dorsata* and *A. mellifera*

nomic analyses (Chap. 5): it can be measured exactly, it is of high heritability, the venation angles are independent of size, and it shows significant geographic variation; this means that it reacts sensitively to processes of evolution even at the subspecies level. In order to estimate the changes in wing pattern numerically, methods of morphometric taxonomy were applied by joint factor analysis of fossil and recent Apidae, using data of 16 wing venation angles (Ruttner et al. 1986). The samples of *A. florea, dorsata* and *cerana* were from Pakistan. Only four complete fore wings of *Synapis* were found among the many specimens investigated. In a second analysis including ten incomplete wings with only four to eight measurable angles, these additional specimens were found to be incorporated in the same cluster, although with greater standard deviation; therefore, the position of the *Synapis* cluster (four wings) can be considered as representative. No attention was paid to the classification of the species into subspecies.

Results: Four compound clusters are visible in the graphic representation of factor 1 and 2 (Fig. 3.8): (a) *Bombus,* (b) *Synapis + A. armbrusteri + A. dorsata,* (c)*A. florea* and (d) *A. cerana + A. mellifera.* In general, the single species are well separated within these groups. The single specimen of *Electrapis* (from Kelner-Pillault 1969) has an isolated position near *Bombus.* The common cluster *Synapis + A. dorsata* shows that wing venation has not changed greatly since the early Miocene. The partial clusters *A. armbrusteri + A. dorsata* (time difference

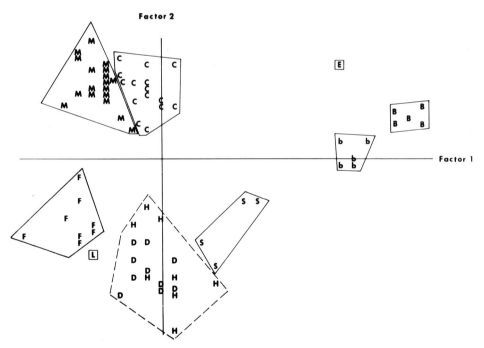

Fig. 3.8 Principal component analysis (PCA) of wing venation pattern in different *Apidae* (16 venation angles). B: *Bombus hortorum;* b: *B. terrestris;* C: *A. cerana;* D: *A. dorsata;* E: *Electrapis;* F: *A. florea;* H: *A. armbrusteri;* L: *A. laboriosa,* M: *A. mellifera;* S: *Synapis*

27

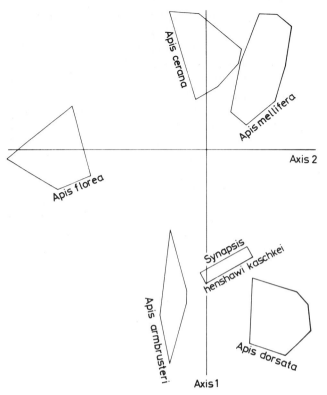

Fig. 3.9 PCA of 16 venation angles, and additionally, length and width of fore wing

10–12 m.y.) are completely overlapping; they are separated only by introducing two parameters of size (length and width of wing), since *A. dorsata* is larger than the *mellifera*-sized *A. armbrusteri* (Fig. 3.9). The similarity of these two types was observed also by Armbruster (1938) and Maa (1953), but now it is evident that *A. dorsata* may be taken as a "mega" edition of *A. armbrusteri*. The clusters of *A. cerana* and *A. mellifera* are positioned at a considerable distance from the *dorsata – armbrusteri* cluster in a different direction.

Only a few fossil honeybees dating from a later epoch have been described. Several specimens from East African copal (allegedly Pleistocene) do not differ in any major aspect from recent *A. mellifera* according to Zeuner et al.(1976). A radiocarbon analysis of this copal by Burleigh and Whalley (1983), however, yielded an age of not more than 100 years.

3.2.3 Estimates of Changes in Speed of Evolution

The evidence of a *dorsata*-type bee in the early Miocene proves the great age of this line of evolution. Since nothing indicates a close relationship of this type with *A. florea*, these two types must have separated already in the Oligocene, which is corroborated also by the large statistical distance *florea – synapsis* (Fig. 3.8). It has to be assumed that the essential *Apis* characteristics, as described in Chapter 1, already existed in this epoch.

Two other estimates of time can be given: The two species *A. cerana* and *A. mellifera* are still in an immature stage of speciation (Ruttner and Maul 1983; p. 150). This indicates a splitting immediately before or during the Pleistocene (about 1–2 m. y. ago). The subunits of the species *A. cerana* and *A. mellifera* are in an initial phase of speciation; they were isolated during the last glaciation and their recent distribution in the temperate zone shows a postglacial pattern. Therefore they have existed only for at most 50,000 years.

Figure 3.10 gives the distribution of centroids of ten geographic races of *A. mellifera* and three races of *A. cerana*. The minimum distance between centroids of *mellifera* and *cerana* (time interval 1–2 m. y.) corresponds closely to the maximum distance of two centroids within *mellifera* (interval 50,000 years = beginning of the last glaciation). In the overall analysis (Fig. 3.8) the clusters *cerana-mellifera* are very close together, but the distance of the cluster *A. armbrusteri + dorsata* to the above cluster is greater than to *Synapis*. No exact estimation of the period can be given as to when the common ancestor of *cerana-mellifera* split from the line *armbrusteri-dorsata*. However, since this cavity nesting *Apis* type was a great evolutionary success, an older age would have had the consequence of rich radiation; therefore, its age certainly must be much less than the interval *Synapis-A. armbrusteri* (12 m. y.).

Morphological invariability indicates a stagnation also in biological evolution.

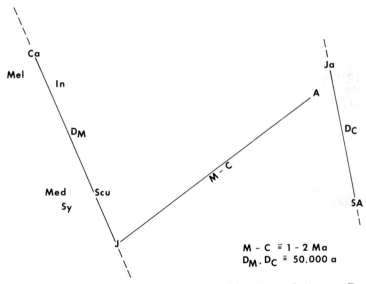

Fig. 3.10 Statistical distances between centroids of races of A. *mellifera*, D_M, and A. *cerana*, D_C (discriminant analysis of 16 venation angles). *Dotted line* distance of extreme samples. M-C: nearest distance between centroids of *A. mellifera* and *A. cerana*. Ca, In, Med, Mel, Scu, Sy, J: races *carnica, intermissa, meda, mellifera, scutellata, syriaca* and *yemenitica* of *A. mellifera*; A, Ja, SA: races *cerana, japonica* and *indica* of *A. cerana*

"The origin of new morphological and biological characteristics and the origin of higher taxa are three . . . different aspects of the same problem". This sentence by E. Mayr (1968) is true also in the reverse sense: unchanged morphology indicates unchanged biology. Therefore it is very likely that the *dorsata*-like *A. armbrusteri* and *Synapis* were tropical open-air-nesting honeybees.

It has to be concluded from these considerations that the two recent open-air-nesting *Apis* species are of very great age. They must have lived through the second half of the Tertiary and the Pleistocene without any essential change and without taxonomic radiation. The only step in this long period of stagnating evolution was an adaptive increase in size, which finally resulted in a giant bee adapted to a temperate climate, *A.(d.)laboriosa*.

As far as wing venation is concerned, an increasingly eruptive evolution is observed in the *cerana-mellifera* group during the Pleistocene. This young radiation was accompanied by the most important change in physiology and behavior since the birth of Apinae namely multi-comb nesting and thermoregulation, and by an enormous extension of its biotope. In this respect, Apinae are paralleled by Bombini, which produced almost 300 closely related species only somewhat earlier and are, therefore, of higher taxonomic rank. But it is peculiar to honeybees that a conservative line of evolution with stagnation since the Miocene and a progressive line with increasing speed of change can coexist. Changes between periods of long-lasting stagnation and rapid evolution are observed also in other animals (Eldredge and Gould 1972; Williamson 1981). In consequence of this particular evolution, the phylogenetic tree of Apinae is a bare structure with a few incipient branches at the extremity (Fig. 3.11) compared to the exuberant ramification of Meliponinae (Fig. 2.3).

3.2.4 Ecology and Historic Zoogeography

Correlating the data on fossil and present honeybees with the geomorphic configuration and climate of Europe and the Mediterranean during the Tertiary period as presented by recent research (Rögl and Steininger 1984; Steininger and Rögl 1985), the outline of a history of honeybee evolution can be attempted.

The fauna of the Baltic amber was tropical and mainly indo-malaysian in composition (Kelner-Pillault 1976). That means that the oriental region extended to northern Europe in the Eocene. *Electrapis* existed in this period, including the possible archetype of Apinae. In the early Miocene (23-24 m.y. ago) the first true Apinae (*Synapis*) appear in central Europe. In the late Miocene (Sarmatian, 12 m.y. ago) when *A. armbrusteri* occurred, the climate was cooler, but still humid and warm (temperate or subtropical). The well-analyzed flora of the Randecker Maar (Gregor 1982), with *Ailanthus, Gleditsia knorii, Embothritis, Ruppia,* etc., indicates an annual mean temperature of 16°C, mean January temperature of 7-8°C and annual rainfall of 1500 mm. Similar vegetation today is found in Florida, ᴜE Brazil, and S China. The reconstructed tropic of this period passed through Europe just north of the Alps. Under these conditions a honeybee species with the same habits as *A. dorsata* (free nesting with a single comb) could have existed.

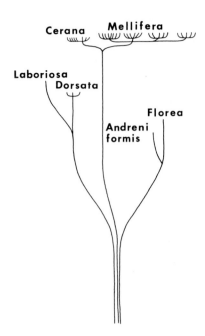

Fig. 3.11 Radiation of the subfamily Apinae

In the Pliocene with a cold-temperate climate no open-air-nesting honeybees could have lived in Europe. Because of their ecological requirements, they must have disappeared from this continent and had a chance to survive only in tropical S Asia, sharing the fate of all subtropical plants and animals of this region. Probably no multi-comb, cavity-nesting honeybees occurred during this epoch in the Mediterranean basin, otherwise they would have been able to survive the Pleistocene as did other adapted thermophile animals and plants.

The puzzling question arises why no tropical Asiatic or European honeybees migrated to tropical Africa, today a paradise for Pleistocenic *A. mellifera* and many (more ancient) Meliponinae. A possible answer is given by the topographic situation of the countries round the Mediterranean during the Tertiary period. In the Eocene and Oligocene, the Mediterranean sea was much wider than today, southern Europe was to a great extent covered by water, the great mountain chains of the Alps, Carpatians, Dinarids, and Apennines emerging as islands ("European Archipelago"). No land connection existed with Africa, the Straits of Gibraltar being wide open and the Indian Ocean extending to the Mediterranean. In the late Miocene the situation was still similar, except for the almost vanished Parathetis (Fig. 3.12). A dramatic interlude during the late Miocene was detected by sea floor research: the Straits of Gibraltar closed, the Mediterranean became a huge dry trough, 4000 m deep and covered with salt ("saline crisis"). A passage existed from the Mediterranean to the Indian Ocean throughout the Miocene, permitting a limited exchange of fauna elements between Eurasia only for short periods, made even more difficult by a permanent desert belt (Rögl and Steininger 1984). The scene changed in the Pliocene to conditions resembling the present: the Mediterranean with the narrow Straits of Gibraltar and a cooler, more humid climate than

31

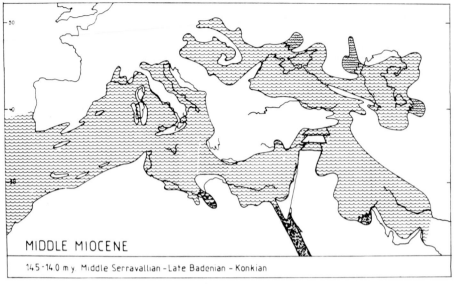

MIDDLE MIOCENE

14.5 - 14.0 m.y. Middle Serravallian - Late Badenian - Konkian

a

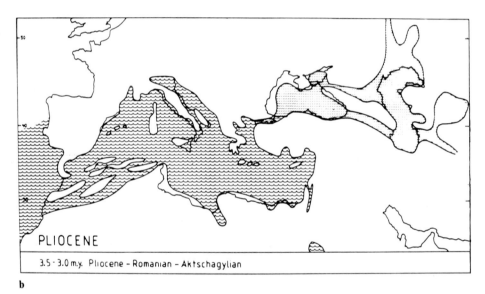

PLIOCENE

3.5 - 3.0 m.y. Pliocene - Romanian - Aktschagylian

b

Fig. 3.12a, b Europe and the Mediterranean during **a)** Middle Miocene, **b)** Pliocene (Rögl and Steininger 1984)

today (Rögl and Steininger 1983, Steininger and Rögl 1985). Therefore, the absence of a Tertiary honeybee in Africa provides further evidence that no cavity-nesting honeybee (which might have been better adapted to migration across adverse zones) existed in the Mediterranean and Middle East before the Pleistocene.

32

3.2.5 The Rise of Temperate Climate Honeybees

Sometime during the Pliocene or early Pleistocene the decisive step must have occurred in the *Apis* evolution: the rise of a perfect thermal homeostasis, making the honeybee largely independent of the environment. The onset of thermoregulation is found in many social Hymenoptera, including Vespoidea, but no single one of them developed the ability to survive several months of low temperature while living in a colony. Even the polytypic Meliponinae with their sophisticated nest covers (Fig. 2.6) never developed this ability.

Single-comb honeybees nest in a "naked" colony, substituting a cover of a thick curtain of bees. This method is as expensive for the energy level as it is inefficient. *A. florea* needs 4.4 times as many workers per comb cell and 3.2 times as many per brood cell as *A. cerana* (Seeley et al. 1982). The brood temperature of *A. dorsata* and *A. florea* is maintained at a fairly constant level during environmental temperatures higher than 20°, but it fluctuates with ambient temperatures at lower levels (Fig. 7.20). To cope with permanent temperatures, such as the prevailing lows in temperate zones during winter, would be completely impossible. The prerequisite of an efficient temperature regulation was the change from open-air nesting to a cave-nesting colony with several combs. This change had to be complemented by a more sophisticated communication in the dark. Perhaps even more important is the behavior of clustering (swarm and winter cluster, protective curtain), typical for all *Apis* species. Apinae represent a "contact type", Meliponinae a "distance type" (Sakagami 1971; p. 2). This perhaps is the reason why the subfamily Apinae exclusively evolved thermohomeostasis.

Several onsets of cavity and multiple-comb nesting are observed in open-air-nesting species: *A. florea* colonies living in caves with broad opening (Fig. 7.16), *dorsata* with multi-comb nests (Fig. 8.12).

Apis dorsata seems to tend in the direction of *Bombus* in its evolution as far as coping with a cool climate is concerned: increased cold resistance by enlarged body size and longer hair. This is seen especially in *A. laboriosa*, "the world's largest honeybee" (Sakagami et al. 1980), which lives in altitudes of 3000 m in the Himalayas with temperatures frequently falling below the freezing point for several hours. To what extent temperature can be regulated in this species has yet to be investigated. A bee of the size of *A. armbrusteri* and *A. cerana* must have started life in hollow trees and closed rock caves during the Pliocene, most likely in a climatic border zone for free-nesting colonies. This is now the case with *A. florea* on the coast of the Persian Gulf (Ruttner et al. 1985a) or for *A. dorsata* in the Himalayas. Considering the distribution of the genus *Apis* at this time, the most likely region for the first steps in a new direction of evolution are the Himalayas. A great variety of differing biotopes close together are always favorable to genetic diversification (see *A. laboriosa*).

From there the new "independent" population spread west and east as far as the climatic conditions permitted. The westernmost region with favorable climatic conditions during even the coldest part of the Pleistocene was the south coast of the Caspian Sea, where many relics of the tertiary survived (but no honeybee of this period). Arid zones of treeless steppes east of the Caspian as they exist also today must have separated the originally unified population into two sections, a western *A. mellifera* and an eastern *A. cerana*. As shown above (p. 33) the separation of the two species must have occurred no earlier than during the Pleistocene. They did not get in touch with each other later because they did not learn to coexist. This hypothesis concerning the region and period of evolution of the *cerana-mellifera* group is supported by several arguments:

1. *A. cerana* evolved sympatrically with the S Asian *Apis* species, while *A. mellifera* did so allopatrically.
2. Tropical Africa was colonized by *A. mellifera* (via Arabia), not by *A. cerana*.
3. The members of the group have only a low taxonomic rank.

The ability of perfect thermoregulation rewarded both species with a huge enlargement of the former purely tropical area of distribution (Fig. 3.13). *A. mellifera* spread very quickly over the whole African continent, retaining there some of the characteristics of its tropical ancestors, such as migrating, absconding, and erratic flight pattern. The tropical and temperate races of *A. mellifera* differ greatly in their behavior, as dramatically demonstrated during their encounter in S America (p. 203). The Mediterranean with all its parts, at this period firmly connected with equatorial Africa by the partly humid Sahara, became the true gene center for *A. mellifera*. Restricted because of the climate during the Pleistocene to the great

Fig. 3.13 The "new" *Apis* territory. Broken line: *A. mellifera*; dotted line: *A. cerana*; comb pattern: tropical *Apis* area

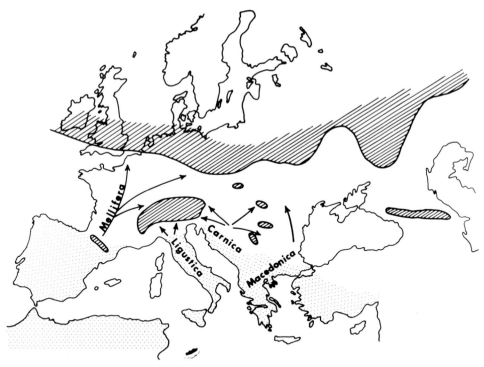

Fig. 3.14 Extension of maximum glaciation, and migration of *A. m. mellifera, carnica*, and *macedonica* to the north and east during the post-Pleistocenic warm period (Atlanticum, 8–10,000 years B. P.)

southern peninsulas, the three important European races originated: *A. m. mellifera* in the Iberic, *A. m. ligustica* in the Apennine, and *A. m. carnica* in the south of the Balkan Peninsulas. A migration of these races to the north and east became possible only in the warm post-glacial period 8–10,000 years ago, together with thermophile deciduous trees such as lime and hazel (Fig. 3.14; Ruttner 1952a). The races of *A. cerana* are of the same age or younger (Chap. 9).

In the *A. mellifera* races of the temperate zone a further important step in evolution took place: the faculty to survive periods of cold without any flight activity for several months. This is not achieved by cavity nesting, temperature regulation, and food storage alone, characteristics also of tropical *mellifera* races. Additionally, the physiologically spezialized type of "winter bees" had to evolve, further formation of a winter cluster, long brood stop close to the end of the winter, blocking of flight activity by cold alone (irrespective of light intensity), a certain resistance to nosema disease etc. (see Chap. 10).

Summarizing the available data, two major steps of evolution of honeybees can be assumed.

A. In the early Tertiary (Oligocene) the principal *Apis*-specific characters originated: a nest in the open consisting of vertical wax combs with bilateral hexagonal cells, protected by a thick cluster of worker bees; progressive feeding of larvae;

division of colonies by swarming with the formation of a temporary cluster prior to colonizing the new nest; communication within the nest by "dance language". This primary *Apis* type was retained throughout the whole Tertiary to the present without substantial morphological and ecological diversification. Confined to the ecological conditions of the tropics, this "conservative" *Apis* type became extinct in Europe as a consequence of climatic deterioration at the end of the Tertiary.

B. In the late Pliocene or early Pleistocene a "progressive" temperate-climate type evolved, characterized by multicomb cavity nesting, temperature homeostasis and elaboration of dance communication. The consequence of these new behavioral characters was a huge enlargement of the *Apis* territory (with recolonization of Europe and colonization of Africa, Fig. 3.13) and rich ecological and morphological diversification at the subspecies level. In both species of this type tropical and temperate races with specific adaptations are found. The high level of general fitness and plasticity of the new type is shown by rapid spread in historical and recent times to various climatic zones of the New World.

Honeybees provide examples of an evolutionary stagnation over an unusually long period as well as of a rapid diversification with incipient stages of speciation.

CHAPTER 4

Geographic Variability

4.1 Morphology

4.1.1 Workers

Qualitative differences in honeybees are restricted only to the rank of species (see Chap. 1, p. 8). The most conspicuously varying character is the male copulatory organ (Fig. 3.5). In general, these species characters show little variation within the species (exception: length of the "indica vein" in *A. cerana*).

On the subspecies level, in contrast, the variability is exclusively quantitative, concerning various parts of the body. The data presented in this chapter are almost entirely from *A. mellifera* and, to some extent, *A. cerana*, since little is known about the other species. Visible differences among domestic bees were noted already in antiquity by Aristotle and Columella (Fraser 1951). These authors described differences in *color* among bee colonies, a character which dominated the systematics of honeybees until now; but color *alone* is frequently inadequate for classification, since its variation is bimodal (p. 69) and not normally distributed.

Differences in *size* can be assessed for taxonomic classification only by measuring isolated parts of the body, using magnifications of $25 \times -50 \times$. The variation between the populations of *A. mellifera* with the smallest and the largest bees were found to range between 25–31% of the smaller value, except for variation in tongue length, which is still more important (Table 4.1). Generally, measurements of size are to some extent correlated with each other, but to a varying degree. Therefore it is quite justified to include several measurements of different body parts to achieve a better discrimination of populations. Alpatov (1929) detected that the body and the appendices (legs, wings, tongue) may vary independently. He described small, but long-limbed and large, but short-limbed bees (Fig. 4.1).

Table 4.1 Total range of variation of several characters in *mellifera* worker bees (mean of colony, mm)

	Proboscis	Fore wing length	width	3rd Leg	Tergite 3+4	Cubital I.	Coverhair
max.	7.26	9.75	3.31	8.61	4.76	3.60	0.13
min.	5.02	7.64	2.64	6.60	3.63	1.53	0.52
Diff.	2.24	2.11	0.67	2.01	1.13	2.07	0.39
%	44.6	27.6	25.4	30.5	31.1	135.0	300.0

Fig. 4.1 Schema of long limbed and short limbed bee. (Alpatov 1929)

Table 4.2 Cell diameter of different races

| Race | Origin | Author | Cell diameter (mm) | |
			Worker	Drone
yemenitica	Oman	Dutton et al. (1981)	4.75	6.20
	Chad	Gadbin et al. (1979)	4.7	
scutellata	Tansania	Smith (1961)	4.8	
monticola	"	"	5.04	
litorea	"	"	4.62	6.15
adansonii?	Angola	Portugal Araujo (1956)	4.8	
"Africanized"	S. America	Rinderer et al. (1986)	5.0	
ligustica	Italy (center)	Alber (1956)	5.27	
mellifera	NW Italy	"	5.37	
carnica	NE Italy	"	5.51	6.91

Size of brood cells is another, more general extrasomatic parameter of worker and drone size. It can be taken as species and as subspecies characteristic (Table 4.2) and is even used as a simple discriminant factor for greatly differing races (e. g., between strains of European and African origin; Rinderer et al. 1986).

Necessarily, most characters used in morphometrics are those of size, and in most programs for morphometric analysis the heaviest load of canonical factors is of this category. In *A. mellifera*, however, many races are accumulated in a very narrow field of size variation (Fig. 4.2). In 11 ponto-mediterranean races (44% of the total) fore wing length varies only within a range of 0.25 mm (difference between means of races). This corresponds to 17.1% of the total variability of the fore wing. It is evident that these races are hard to discriminate exclusively by size.

On the other hand, several races of *A. mellifera* can be completely separated by a few characters of size alone, as shown in Fig. 4.3 for the races *yemenitica*, *meda*, and *mellifera*. Size variation is only slightly less in *A. cerana* than in *A. mellifera*. Hind leg length of the smallest bee sample is 84.5% of the largest in *cerana* and 76.6% in *mellifera*; the corresponding values for fore wing length are 84.0% and 80.2%, for hair length 39.9% and 28.7%. The minima are almost equal, but the maxima are higher in *mellifera*, as shown in Fig. 4.4. The smaller range of variability in *A. cerana* can perhaps be explained by the fact that only 93 samples of this species were included in the analysis compared to more than 800 in *A. mellifera*.

The introduction of a new category of characters, namely the pattern of wing venation, marked a definite step forward. At first, relations between certain venation segments were used (e. g., cubital index, CI; Goetze 1930, 1964); later, angles formed by connection lines between venation crossing points were added

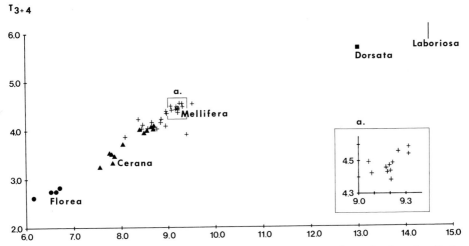

Fig. 4.2 Variation of body and wing length (*T3 + 4, FL*) in *Apis* species and subspecies. Each point represents a group mean [or a single sample in *A. dorsata* and *A.(laboriosa)*]. Overlapping occurs only with *A. cerana* and *A. mellifera*. Insert *a* = 11 mediterranean *mellifera* races of almost identical size

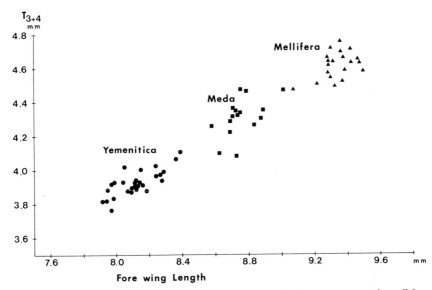

Fig. 4.3 Morphometric discrimination without overlapping of the *yemenitica-meda-mellifera* groups of *A. mellifera* by only two characters of size (fore wing length and T_{3+4}). Each point represents one sample. *Yemenitica* = Yemen and Oman, n = 21; *meda* = Iraq, n = 16; *mellifera* = northern Europe, n = 10

39

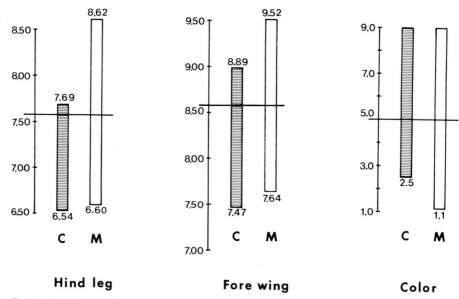

Fig. 4.4 Minimum and maximum values of characters of size (hind leg, fore wing) and color (T₃) in *A. cerana* (*C*) and *A. mellifera* (*M*). *Horizontal bars* median of *A. mellifera* values. (Ruttner 1984).

(DuPraw 1964, 1965 a, b). This new set of quantitative characters, not correlated with size, shows normal distribution and evidently has no adaptive value. Using wing venation alone (or combined with measurements of wing size) a fairly good intra-specific classification of *A. mellifera* was achieved (DuPraw 1965 b) as well as a new interpretation of fossil bees from the Miocene (Ruttner et al. 1986). However, reducing morphometrics to characters of the fore wing means giving away a wealth of potential taxonomic information, as shown by comparison of DuPraw's results with the present state of taxonomic analysis.

Another independent category of characters comprises data on the hair cover of worker bees (length of cover hair, width of tomenta; Fig. 6.2) which show great geographic variability. Using one single character, length of abdominal cover hair, two races, *mellifera* and *carnica*, show total separation without overlapping. By adding a second character (CI) even a slight degree of hybridization can be detected (Ruttner 1983).

From the examples cited it becomes evident that several character categories have to be used in honeybee morphometrics to obtain consistent results with optimized discrimination. Only in populations with great statistical distances, such as *mellifera-carnica* or European and tropical African races, may a few, properly selected characters be sufficient. However, in a set of only a few, a single biasing character may completely distort the result. In an analysis of Mediterranean races, color, one of only six characters used, forced all races into only two groups, one "yellow", the other "dark" (Cornuet et al. 1975). On removing color from the analysis the remaining five characters gave a plausible segregation into the expected number of geographic populations.

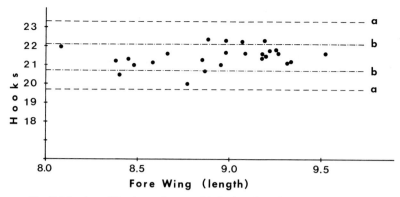

Fig. 4.5 Geographic distribution of hook numbers on hind wing. Standard deviation (*a*) within, (*b*) between subspecies. No. of hooks plottet against wing length. Each black dot represents the mean of a geographic *mellifera* race

It is by no means certain that the 40 characters used in our investigation are the best and most complete selection. All of these characters are the result of thorough screening, some of them since the early days of honeybee morphometrics. Nevertheless, this did not prevent some of them, such as for example the frequently used number of hooks on the hind wing, from finally proving to be of no use for taxonomic discrimination: no clear geographic variation was found in the 25 groups from the whole *mellifera* area (474 samples with almost 9000 bees) in the number of hamuli, and the within-variability was more than threefold the between-variability (1.835 vs. 0.588; Fig. 4.5). This is surprising considering the high heritability of the character, (0.68; Oldroyd and Moran 1983) and the successful selection of a "low" and a "high" line out of a *carnica* strain (Goncalves 1979), with $\bar{x} = 10.6$ (s. d. 1.07) for "low" and $\bar{x} = 28.6$ (s. d. = 1.97) for "high" – that is a larger range than observed naturally in the whole genus *Apis*! The situation is similar in *A. cerana*; the mean number of hamuli of 41 samples originating from nine different regions was 18.22 (min. 17.68 in Himalaya, max. 18.85 in Pakistan). Average standard deviation of group means was 0.395, the mean s. d. of single groups was 0.539. (An exception are perhaps the bees of the Philippines: $\bar{x} = 16.62$, s. d. 1.52; 4 samples). Only now do we dare to eliminate this traditional character of honeybee taxonomy (see Alpatov 1948, Goetze 1964), which is still widely used (Daly and Balling 1978; Mattu and L. R. Verma 1980; 1984b; Rashad and El Sarrag 1984).

On the other hand, special characters not pertaining to the set of "standard biometry" have to be added for clear description of some subspecies. *A. m. capensis*, for instance, is difficult to discriminate from *A. m. scutellata* by the usual methods. A discrimination is, however, easily achieved if two characteristics reflecting the peculiarity of reproduction (thelytokous parthenogenesis) are included, namely diameter of spermatheca and number of ovarioles (Ruttner 1977b; Moritz and Kauhausen 1984; Crewe 1984).

A special problem are derived characters as, e. g., indices or sums, as already introduced by Alpatow (1929), to describe the morphometric peculiarities of a taxonomic unit. Indices are problematic in statistical analysis (Daly 1985), and this is

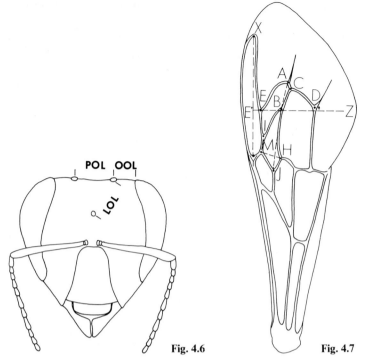

Fig. 4.6   Position of ocelli (Maa 1953). *POL* post-ocellar line; *OOL* oculo-ocellar line; *LOL* lateral ocellar line

Fig. 4.7   Indices of wing venation (Goetze 1964). *CB:BA* (=a: b) cubital index; *BJ:MH* precubital index; *BA:IE* dumb-bell index; *E'X:E'Y* radial index; position of point *D* relative to vertical line *E'Z* discoidal index

why also in this study only primary data are used. The abstract statistical functions provide discrimination and distances as indications of similarities. For presentation of a biological individuality, for a clear physical, scientific description of a taxonomic type, secondary data provide important elements: the sum of the single joints gives the total length of the hind leg with accumulation of slight single differences; the index of sternite 6 quantifies the abdominal proportion ("index of slenderness", S6I); the hind leg may be long, compared to other races, but short in relation to the body size of the same individual as, for example, in *A.m.mellifera*; the cubital index (CI) reflects in one figure the whole apical part of the fore wing venation pattern. In the special chapters the definition and discrimination of races will be based on primary data, the description, however, partly on secondary values.

It is evident from the few examples given above that "the classification of the taxonomic categories below the rank of species ought to be made only on the basis of a detailed and thorough investigation of the geographic variation of a given species" (Alpatov 1929, p.47).

A number of measurements were used in honeybee morphometrics besides those described in Chapter 6: width of head and thorax (Goetze 1930), interocellar and oculo-ocellar distance (Fig. 4.6; Maa 1953), length of antenna or its segments, distance and diameter of eyes (Stort 1979; Rashad and El Sarrag 1980), length and width of hind wing (Daly et al. 1982), density of spines on the wing surface (Woyke 1976), length of venom and rectal glands (Firsow 1976). Besides its discriminatory potential the exactness of measurements and the amount of labor involved, as well as other aspects, have to be considered prior to inclusion of a character into a large-scale morphometric program.

Several indices of wing venation were introduced especially for discrimination of *A. m. mellifera* from other races (Fig. 4.7; Goetze 1964; Louis 1963): cubital index (CI), precubital index (length: width of cubital cell 2), dumb-bell index (B-A: I-E), radial index (E'-X: E'-Y), discoidal index (relative position of point D to the prolongation of line E'-E). All of these indices, except for the very important CI, became obsolete since the introduction of venation angles.

4.1.2 Drones

One hundred and sixty drone samples (with 20 individuals each) from the majority of the races were included in our morphometric analysis. From the following groups no drones were available: *A. m. cecropia, cypria, litorea,* and *syriaca.* Variation is evidently of the same amplitude among drones as among workers (Table 4.3). A correlation exists between the two castes: in colonies with large workers large drones can be expected. In general, also the geographic variability is identical; therefore, the drone phenogram of *A. mellifera* is very similar to the worker phenogram, although some deviations are observed (Fig. 10.8). The phenetic structure of the species *A. mellifera* is even clearer in a drone than in a worker phenogram: on axis 1 of a PCA *yemenitica* and *adansonii* at the minus pole and *mellifera + caucasica* and *carnica* at the plus pole; *meda* at the minus pole of axis 2 and *adami + unicolor* at the plus pole (Fig. 10.8). As far as color of the exoskeleton is concerned, drones in the majority of the races tend to be darker than workers; in several cases (e.g., *adami, anatoliaca, adansonii*) colonies with yellow workers have completely dark drones. Pigmentation of hair on the thorax may be very characteristic for certain races although it can hardly be exactly measured: tangrey in *carnica*, brown or black in *mellifera*, intense black in *caucasica*, yellow in *ligustica*. This character is used for discrimination between races using color plates (Ruttner 1983).

Table 4.3 Maximum and minimum values of several characters in *A. mellifera* drones. Measurements of length in mm

	Fore Wing Length	Width	Length Tibia	Metatars.	Cub. Vein A	Angles A4	D7	Color T2
Max.	12.693	4.220	4.193	2.753	6.495	36.74	119.60	8.15
Min.	10.459	3.311	3.315	2.165	3.540	24.30	101.65	0.0
Diff.	2.234	0.909	0.878	0.588	2.955	12.44	17.95	
%	21.4	27.5	26.5	27.2	83.5	51.2	17.66	

4.2 Biochemical Variation

Mestriner (1969) and Sylvester (1976) started the study of enzyme polymorphism in honeybees. After the investigations were extended to autochthonous strains in Europe and N Africa, valuable progress was achieved in the knowledge of geographic variability of honeybee enzyme patterns. At present, the following polymorphic enzyme systems are known (Sheppard and McPheron 1986): malate dehydrogenase (Mdh-1, three or four alleles), esterase (Est, three alleles), phosphoglucomutase (Pgm, two alleles), malic enzyme (Me, two alleles) and aconitase (Acon-2, two alleles). Further, polymorphism was reported from S America for alcoholdehydrogenase (Adh-1, two alleles) and a general protein (P-3, two alleles; Sylvester 1985).

Data on allele frequencies of Mdh-1 and Est yielded clear differences between *A. m. mellifera* in France and *ligustica* (Badino et al. 1982; Cornuet 1983), between *ligustica* and *sicula* (Badino et al. 1985) and *carnica* and its hybrids in CSSR (Sheppard and McPheron 1986). Analyzing Mdh-1, Adh-1, and P-3 in the same sample, Sylvester (1982) was able to detect hybridization with Africanized bees with a probability of 0.01.

As a prime example a fairly complete analysis of the Mdh pattern of the Western Mediterranean is available, revealing a striking conformity with the morphometric results (Fig. 4.8): transitions from *intermissa* to *sicula* and *iberica*; hybridization of *ligustica* with *mellifera* and *sicula*; close relation of *iberica-mellifera* and *ligustica-carnica*.

Another possible source of biochemical variation are hydrocarbons in the surface cuticular waxes. Substantial differences in composition and quantity were found among different *Apis* species and to some degree between S African (and

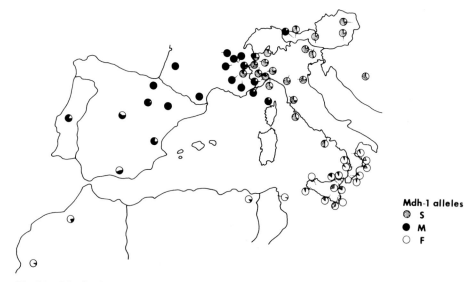

Fig. 4.8 Distribution of Mdh-1 alleles in the western Mediterranean. (Cornuet 1982b, 1983; Badino et al. 1984)

Africanized) and European bees (Francis et al. 1985). It has yet to be shown, however, whether the geographic variability is influenced by environmental factors and to what extent it can be considered subspecies-specific (MS Blum 1986).

4.3 Geographic Variation in Behavior and Physiology

Ample evidence exists from long-standing apicultural practice that there are important behavioral differences between geographic races. They are described in more detail in the appropriate chapters (10-14); here only a few examples: Great differences in defense behavior and irritability ("nervousness") exist within European races and even more between African and European races (Ruttner 1975; Stort 1974). The average number of swarms per colony may vary for a factor of more than 10 between European and tropical populations (Winston et al. 1983). The brood rhythm varies basically among races of the Mediterranean basin *(ligustica, sicula, intermissa)* and *A. m. mellifera* from Central Europe (Chap. 10; Lunder 1953; Ruttner 1975), but also among populations of the same race (Louveaux 1969). The use of propolis as construction material is as variable as the robbing tendency. Differences between *ligustica* and *carnica* in the way of orientation, observed by apiculturists keeping track of the amount of drifting, were analyzed experimentally by Lauer and Lindauer (1971, 1973) and by Riecke-Lauer and Lindauer (1985). Given the choice at the feeding place, *carnica* bees orient themselves primarily by environmental structures and react strongly to a changed position of the feeding dish. *Ligustica* quickly become used to a displacement of the dish but always react to color. Analogous experiences are made in apiculture: *carnica* is very accurate in homing when compared with *mellifera*, with little drifting to neighboring colonies (Ruttner 1954a). This behavior might be of advantage when small beehives are piled up in stacks with many entrances close together in one block, as in Slovenia. In contrast, if a single *ligustica* colony is placed in the midst of a row of ten *carnica* colonies, yellow flight bees will soon be seen in every other hive and the *ligustica* colony will lose in strength, since no drifting grey bees compensate for the loss.

Another aspect of this behavioral variation are differences in flower constancy. Gasanov (1967) tested four races (*ligustica, caucasica, carnica, mellifera*) on various simultaneously flowering plants. *Mellifera* showed the highest and *ligustica* the lowest flower constancy (=0.8% and 14.4% respectively changed the flower type during one flight), *carnica* and *caucasica* were intermediate (3.8% and 9.1% changes respectively). Ursu (1976) stressed that both flower constancy and flower migration might have advantages and disadvantages for the bee colony. Constancy is a more efficient collecting strategy but poses the risk of fixation on a pollen source of minor quality. Migration may also result in lower quantity of pollen load, but higher quality because of greater diversification of the components (especially of essential amino acids).

The frequency of waggle circuits indicating a certain distance is strictly subspecies-specific (Fig. 4.9) and evidently correlated with preferred foraging and dispersal distance: "slow dancers" seem to be short-distance foragers and swarmers and vice versa (Gould 1982). Several components of the dancing behavior, besides

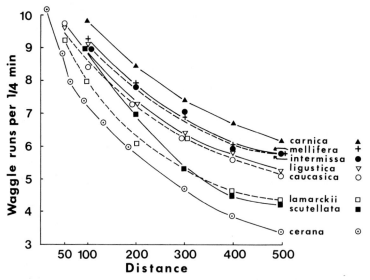

Fig. 4.9 Different dance rhythms in various races of *A. mellifera* and in *A. cerana* from Sri Lanka. (v. Frisch 1967)

speed, are variable among races: the duration of the acoustic signal during the waggle run is shorter in *carnica* than in *ligustica* (434 ms vs. 635 ms at 200 m) and the number of pulses lower (12.9 vs. 20.9; Eskov 1976). Finally, distinct differences in learning ability among races were found (Koltermann 1973b; Menzel et al. 1973; Brandes 1984).

In a number of cases the adaptive significance of behavioral peculiarities is evident. This occurs in consequence of long-lasting evolutionary selection: European strains have retained their low swarming rate in tropical S America through centuries, although this renders them completely inferior in fitness, as was shown when the Africanized bee appeared as competitor. Transferred colonies usually maintain all their inherited behavioral inventory and all reports on "acclimatization" refer to nothing else but hybridization. Defining the taxonomic unit to which a colony belongs allows also for behavior prediction.

Laying Workers. The reproductive disposition of worker bees, an important phenomenon in the social organization of honeybees, shows a significant variation between and within *Apis* species. It can be measured by the interval between dequeening and start of egg laying by worker bees ("latency period"), by the number of worker ovarioles, and by the number of eggs laid by workers within a given period.

These parameters were experimentally investigated in seven *mellifera* races (Ruttner and Hesse 1979). The latency period varied between 5.6 days in *intermissa* and 30 days in *carnica* (Fig. 4.10). European races had a long and African races a short period (according to Buttel-Reepen 1921, in *A. m. lamarckii* workers start egg laying within a few days after dequeening). The number of eggs laid within 48 h was greatest in *adami* (655) and lowest in *mellifera* (27.7). Number of

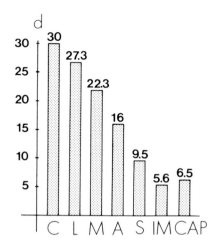

Fig. 4.10 Average latency period of laying workers in various races of bees (days). C: *carnica*; L: *ligustica*; M: *mellifera*; A: *adami*; S: *scutellata*; IM: *intermissa*; CAP: *capensis*. (Ruttner and Hesse 1979).

ovarioles was 3.21, 3.26, and 3.73 in *mellifera, carnica,* and *ligustica* respectively, but 7.1, 7.4, and 9.4 in *scutellata, intermissa* and *capensis.* Latency period and number of eggs were negatively correlated.

4.4 Genetics of Subspecies Characters

Everybody expects bees to maintain all their original characteristics if transferred to a completely new environment. This was proven scientifically by Alpatov (1929) by measuring samples of yellow bees from different localities in N America, with the result that no substantial differences existed with original *ligustica* bees from Italy. However, there was less pigmentation on the abdomen of American Italians – undoubtedly an effect of the color preference by beekeepers. Another such similarity was found between black bees from Ontario, North Carolina, and Florida, and *mellifera* from Europe. Ruttner (1978) compared isolated Italians from Kangaroo Island (Australia) and Black Bees from Tasmania, with exactly known importation date (1825 and 1835, respectively), with data on *ligustica* and *mellifera* from Europe. No significant difference was found, and *mellifera* from Tasmania turned out to be less hybridized than their parents in Britain.

The inheritance of color characteristics was studied by Roberts and Mackensen (1951). They found a continuous color variation from dark to yellow in the F_2, and postulated a series of seven genes responsible for the color difference between their yellow and black strain. A polygenic inheritance was found also for the CI in crosses of *carnica* and *mellifera* (Ruttner 1953).

Heritability is high for morphological characters, as was to be expected from these facts. Moritz and Klepsch (1985) calculated high h values for different wing characteristics (length, width, five venation angles), applying a nested breeding scheme with parthenogenetically reproducing *capensis* workers. Similar results were obtained by Oldroyd and Moran (1983) for the heritability of the number of wing hooks. Heritability of learning behavior is much lower (Brandes and Moritz 1983) and compares to other behavioral characteristics like honey yield, for example.

Table 4.4 Frequency of queenless nuclei (*capensis/carnica × capensis*) with a varying amount of worker offspring (0–100%) in two consecutive seasons (unpubl. data)

Year	n	0–10	11–20	21–30	31–40	41–50	51–60	61–70	71–80	81–90	91–100
						% of thelytokous workers					
1978	53	16	2	3	1	–	–	–	1	4	26
1979	57	21	3	3	1	1	1	1	1	4	21
Sum	110	37	5	6	2	1	1	1	2	8	47

In contrast to these apparently general multifactorial genetic systems for subspecies characters, one of the most unique peculiarities of *mellifera* races, thelytokous parthenogenesis in *A. m. capensis*, is caused by two alleles of a single gene. The character "diploid eggs laid by workers" of the Cape bee is recessive in a cross of *capensis × carnica* (dominance reported by Ruttner in 1977 in a preliminary study later proved to be an error of method, since in this first experiment drifted pure *capensis* workers had evidently entered the hybrid colonies). To study the genetics of this phenomenon, pure *capensis* queens were instrumentally inseminated with the semen of one single segregated drone each. These drones were produced by F_1 queens of the cross *capensis × carnica* and represent genetically male F_2 gametes. The inseminated queens were removed from their nuclei prior to the hatching of their brood. Then the start of worker oviposition in the queenless colonies was recorded and the sex of the worker's offspring was determined after pupation (a total of 110 nuclei in 2 successive years, (Table 4.4).

The segregation was strongly bimodal, suggesting two alleles of one gene influenced by some side effects. These are probably the following:

1. Thelytokous parthenogenesis occurs in a low percentage (1%) also in European races (Mackensen 1943) and *capensis* workers may also produce some haploid eggs. Therefore, the manifestation of the alleles in question is not total and the two classes expected in monofactorial inheritance are not 0 and 100%, but 0–10 and 90–100%.
2. Segregated parthenogenetic workers may have drifted to other nuclei in spite of the well-spaced arrangement of the boxes.

Therefore, the results obtained, identical in 2 years and representative as far as the number of units are concerned, may be regarded as the best approximation which can be expected (Table 4.4). No significant difference exists between the two extreme classes.

4.5 Effects of Environment

Michailov (1927 a, b, c) studied the influence of single environmental factors on a number of morphometric characters (length of tongue, hind wing, tergite 3 + 4, width of hind wing, number of hamuli on hind wing; review Alpatov 1929). The investigations were performed at the Tula Experimental Station (about 165 km south of Moscow); thus it can be concluded that bees of the race *A. m. mellifera* were used. These fundamental investigations were confirmed later by authors working with other bees and also with some other characters. Of the many factors studied, four groups may be of significance for research on geographic variability:

4.5.1 Seasonal Variability

Samples of young bees were taken on 5 days between early June and early September. A significant increase in size of the characters was found between sample 1 and 3 (during the swarming period, June 6 to July 20) of 2.6–4.5%, while later in summer the values remained unchanged (Michailov 1927a). Mattu and Verma (1984a), measuring 39 characters of *A. cerana* from North India reported a similar trend in 17 of these characters, although the others did not show a significant difference. The most important increase (2–3%) during the course of the early season was found in the length of sternites 3 and 6, tergite 3 and antenna.

4.5.2 Size of Brood Cell

Bees of a given race construct combs with mostly constant, characteristic cell diameter, highly correlated with body size of adults (see Table 4.2). If bees are forced to rear brood in cells of a size other than the specific one – smaller or larger –, then the size of emerging bees is changed correspondingly. Michailov (1927c) compared two groups of worker bees from one and the same colony, the one being reared in worker cells ($\varnothing$ 5.0 mm, depth 11.4 mm), the other in drone cells ($\varnothing$ 6.96 mm, depth 12.55 mm); this is an extreme difference, and the queen will only exceptionally lay fertilized eggs into cells of such a large diameter. The worker bees reared in drone cells were significantly heavier (0.145 vs. 0.130 mg = +11.45%) and larger (tongue length +4.82%, length of fore wing +2.69%) than the bees from normal worker cells. The characters of workers from large cells showed a higher standard deviation. Worker bees of *A. cerana* react in the same way (longer proboscis) to larger brood cells (Jagannadham and Goyal 1980). These experiments were repeated later on (review Grout 1963) with the same positive result. The final goal of this work, however, to increase honey yields by producing long tongued bees, was not achieved.

For sampling bees for morphometric analysis the opposite situation is more important: bees reared in old, dark combs with reduced diameter. Michailov (1927a) compared bees reared in dark combs (17 and 21 brood cycles, respectively) with bees reared in white combs, with both kinds of comb simultaneously in the same colony. Bees reared in the dark combs were slightly, but significantly, smaller than those in the light combs, provided the difference in cell diameter was at least 5–6%. These results were confirmed by several authors. In extreme cases (combs with 68 brood cycles) the worker weight was reduced by 18.8% (reviews Grout 1963; Zander and Weiss 1964). Therefore, the deviation of single samples can perhaps be explained by this environmental factor. It has to be noted, however, that naturally constructed cells are far from being equal in size (Hepburn 1986).

4.5.3 Quantity and Quality of Brood Food

Larvae fed in strong colonies develop into slightly heavier bees with longer proboscis than larvae reared in weak colonies (coefficient of correlation r = 0.59 and 0.49, respectively; Alpatov 1929). Furthermore, it seems likely that seasonal variations in size of bees are basically caused by the quantity of brood food deposited in the cells.

An interesting phenomenon is the repeatedly reported influence of nurse bees of different origin on morphological characters of bees reared by them (Michailov 1928, Ruttner 1957a, Smaragdova 1960, Rinderer et al. 1985). The characters of the bees are slightly shifted in direction of the characters of the nurse colony. Since characters of wing venation (besides characters of size) are also influenced (Ruttner 1957), these results cannot be merely the consequence of differences in food quantity. Indeed, qualitative differences were found in the brood food of *A. m. mellifera* and *A. m. caucasica* (Smaragdova 1963).

The question of a specific influence of nurse bees on characters of the reared bees has a basic theoretical, even ideological background. In the years 1955-1970 it was claimed that the influence of the nurse bees is not merely phenetical, but also genetical – a contention which was not confirmed by later experiments (for discussion of this problem see K. Weiss in Ruttner 1983). Another possible non-genetic factor acting on wing venation is a certain oscillation of development homeostasis as shown by some nonsymmetric structures. If the left and right wing of individual bees is compared, a significant dyssymmetry in cubital index is found. The mean of large samples (20-100 bees), however, is identical for both sides (Ruttner 1952b).

4.6 Sensitivity to Diseases

In apicultural literature many reports are found regarding differences in susceptibility to the various diseases among the races. Br. Adam repeatedly mentions the high susceptibility of *ligustica* strains to American foul brood and the relative resistance to all brood diseases in *carnica*. Sturges (1928) found lower infestation and fewer losses through acariosis in Italians during a severe epidemic in England in the 1920's than in the local Black bee. This observation was repeatedly confirmed since then by Br. Adam (1983).

In other cases imported races are more susceptible, as was found, for example, regarding Nosema disease in *caucasica* in cool climates (pers. unpubl. data). This points to a problem of general significance: many infectious diseases become manifest only if the colony lives in a state of unbalance. What appears to be "high susceptibility" of a strain is frequently nothing but a lack of adaptation to an inappropriate environment. However, since the classic investigations of Rothenbuhler (1964), it is beyond doubt that a true resistance against AFB exists. Experience in flight rooms showed that *A. cerana* is resistant to Nosema disease (Chap. 9) and Awetisjan and Maximenko (1979) and Sidorov (1969) found differences in Nosema susceptibility in a number of strains and races of *A. mellifera*.

4.7 Patterns of Geographic Variability

One of the first accomplishments of honeybee morphometrics was the detection of a north-south cline in length of proboscis. Several authors (Cochlov 1916; Michailov 1924; Alpatov 1925; Skorikov 1929 a) found during their studies of honeybees of the Russian plains that the length of proboscis increases gradually from north to south (Table. 4.5).

In addition to this graded variation of tongue length, the hind legs also show the same tendency of increasing length to the south, whereas the entire body becomes substantially smaller from north to south in the same area (Alpatov 1929). As a consequence the body proportions shift from a big, short-legged to a smaller, long legged bee (Fig. 4.1). This seemed to fit very well to Bergmann's rule and Allen's rule: geographic races of one species are larger in the north than in the south, and appendages of the body (leg, wing, beak) are relatively shorter in the north than in the south. The same gradual variations were found also with increasing altitude. Described primarily for vertebrates, these observations on insects received much attention and are cited in textbooks to date.

Unfortunately, these data do not actually prove what they were supposed to: the existence of a gradation of quantitative characters parallel to a gradation of climatic conditions. As later studies have shown (Ruttner 1952, Goetze 1964), the gradation observed is partly a mere consequence of the hybridization of the short tongued and large *A. m. mellifera* from western and northern Europe with the long-tongued, smaller *A. m. carnica* (and *macedonica*) from south east (Fig. 4.11). In Central Europe both races are sharply separated by the Alps, while in the eastern plains a smooth transition occurs from the "Forest bee" (A. m. silvarum = *A. m. mellifera*) of the north to the "Steppe bee" (A. m. acervorum = *A. m. macedonica*). of the south. Thus the gradual variation of tongue length from Leningrad to Krasnodar has to be listed, at least partly, in Huxley's group of "grade characters not adaptively correlated with environment".

If the first attempt to proof a true climate-correlated variation in honeybees did not hold up to a re-examination on a larger scale, this does not exclude that gradual geographic variations exist elsewhere in *A. mellifera*. Scrutinizing the honeybee data bank of the Apicultural Institute, Oberursel, a graded north-south variation was found extending along the countries of the west coast of the Atlantic ocean

Table 4.5 Tongue length in relation to geographic latitude in the USSR. (Skorikov 1929 a)

Location	Degree N latitude	Tongue length (mm)	Difference
Leningrad	60	5.73	
	59	5.95	0.22 = 3.84%/1° lat.
	57.5	6.00	
Moskwa	56	6.12	0.08 = 1.33%/1° lat.
Tula	54	6.22	0.05
	52	6.28	0.03 = 0.56%/1° lat.
	48.3	6.43	0.04
Kuban	46	6.53	0.04
Krasnodar	45	6.54	0.01 = 0.15%/1° lat.

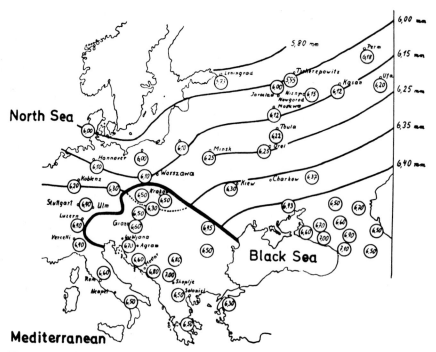

Fig. 4.11 Isoglossae in Central Europe and border line between *A. m. mellifera/carnica* (thick bent line from the Adriatic to the Black Sea; Goetze 1964)

(Figs. 4.12, 4.13; Ruttner 1985a). Throughout this "Atlantic chain", extending from Scandinavia to the Cape of Good Hope (94° of geographic latitude), a gradual increase or decrease of quantitative characters can be established. This is probably one of the longest continuous chains of gradual variations so far found in animals.

The decrease in size from high to low geographic latitudes concerns all measurements taken: sclerites of the ventral and dorsal side of the abdomen, length of hair (Fig. 4.12), forewing, proboscis and third leg. It is important to notice that the tendency of variation is reversed as the geographic latitude increases south of the equator (though this effect is somewhat obscured by the fact, that a sample of 20° S latitude comes from the Namibian desert, the location with the highest temperatures and therefore with the smallest bees of the series). Thus highly significant correlations are found between measurements of body size and geographic latitude (r = 0.80 to 0.885). Special consideration needs to be given to the variation of length of the proboscis. This body part varies strictly correlated with the other measurements, from the southernmost location (Cape Town) to 40° N latitude (southern France). From there up to the northern boundary of the distribution of the species (Oslo and Leningrad, 60° N latitude) a dissociation is observed: the length of the proboscis decreases, with increasing body size (Fig. 4.13). The same phenomenon is observed in the length of the 3d leg. In consequence the coefficient of correlation with geographic latitude is considerably lower for these two characters (0.63 and 0.72 respectively).

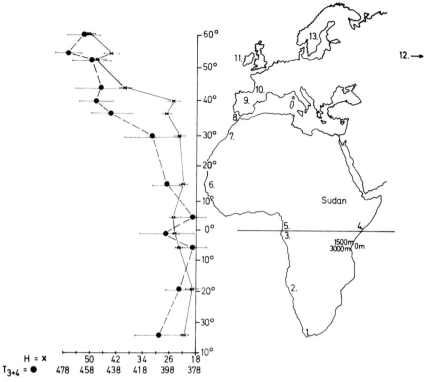

Fig. 4.12 Graded variation of hair length (*H*) and tergite width (*T₃*) along the Atlantic coast. No. *1-13*: sampling locations. (Ruttner 1985 a)

This gradual change to a large, short tongued, short legged and long haired bee is observed within the area of the West European Black bee (*A.m.mellifera*) only. Thus the "isoglossae" of Alpatov and others as a true geographic variation (geocline) are confirmed also for western Europe. However, this term can be applied in this case (diminuation of length of tongue and legs) only for this single race, which extends farthest north. This special trend, quite different from the other races of the Atlantic chain, seems to confirm Allen's rule also for an insect. However, from the tropics up to the warm temperate zone, legs and proboscis vary conformingly to the body. It has to be stressed that the fore wings continue to increase in length towards the north (Fig.4.13); these are organs which have little cooling but, in contrast, rather a protecting effect on the body.

Along the coast of the Atlantic, the index of st6 (Chap.6, Fig.6.7) varies along with the body dimensions: bees of low latitudes have a slender abdomen (index 86), those of high latitudes a broad abdomen (index 76). A coefficient of correlation of 0.885 was found with the geographic latitude. The third group of characters with directed variation concerns the hair: bees of cooler climate have longer hair, those of warmer zones of the Atlantic coast, shorter hair (Fig.4.12). This corresponds to observations in mammals (rule of Rensch 1936)

As far as color is concerned, a kind of reversion of Gloger's rule can be

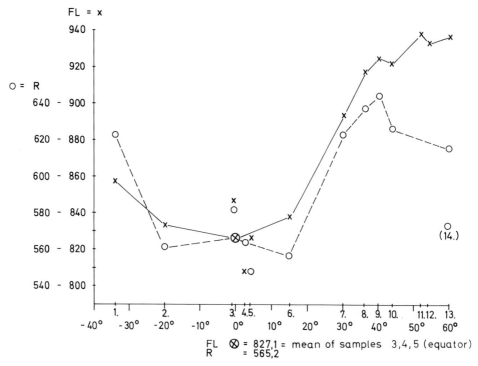

Fig. 4.13 Graded variation of length of fore wing (*FL*) and proboscis (*R*). (Ruttner 1985a)

observed. This rule says that within species of warm-blooded animals the melanin content increases towards warmer zones. In honeybees, in general, races of cool climates are more pigmented than those of warm climates. Abrupt changes are found in this character instead of gradual transitions in pigmentation. In the Atlantic chain of races, types with dark pigmentation are found at both ends, in zones of high geographic latitude: *mellifera, iberica* and *intermissa* in the north, *capensis* in the south. The same holds true for a race of higher altitudes, *A.m.monticola*. The bees found on both sides of the equator show only little pigmentation.

To this horizontal, long-distance geocline corresponds a vertical, short-distance "ecocline". As shown by Smith (1961) in East Africa (Tansania), three different races are found within a distance of only 300 km and a difference in altitude of 3000 m: *A.m.litorea* in the coastal zone, *A.m.scutellata* in the savanna (1000–2000 m) and *A.m.monticola* up to 3000 m in the Rain Forest of the Kilimanjaro.

A gradual variation correlated to altitude is observed, including all the characters varying in horizontal direction, with one exception: the proportions of abdomen (index of slenderness) remain the same irrespective of body size (Fig.4.14; Ruttner 1985a).

It has to be emphasized, that the lowest values of characters of body size and hair length do not necessarily occur at point 0.0° of latitude, but in zones with the highest temperatures (mostly inland, as in Sudan or on the Arabian Peninsula). On

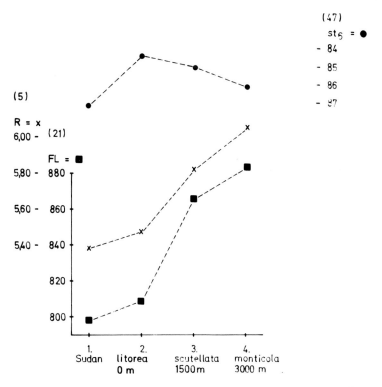

Fig. 4.14 Ecocline from Sudan and the coast to 3000 m (Kilimanjaro) in East Africa. *FL* fore wing length, *R* proboscis, *St₆* index of sternite 6 (slenderness). (Ruttner 1985a)

the other hand, the largest body is found in the Ural mountains. This indicates that the factor of environment mainly correlated with body dimensions is temperature. Since samples are available only from a restricted number of locations along the total gradation, it cannot be said whether or not the inclination is uniform along the whole area. But several geographic races are involved in the long-distance geocline and also in the ecocline. Therefore, this type of graded variation can be designated as "discontinuous stepped cline" according to the terminology of H. Huxley (1939).

These general rules of variation as shown in the races of *A. mellifera* along the eastern border line of the Atlantic are found in some way also in the three other *Apis* species (as far as measurements were taken). In *A. cerana*, the northern hill variety is larger and darker than the variety of the plains and the islands in the south. In *A. florea* the bees of the north (Pakistan, Iran) are distinctly larger than those of the south, e.g., Sri Lanka. *A. (laboriosa)* from the Himalayan mountains, in altitudes of 3000m, belonging to the *dorsata* group, is the largest honeybee observed so far.

These close relations between morphometric variation and geographic latitude in one line of races along the Atlantic and similar observations in the other species, however, cannot be generalized. Morphometric variation of honeybees is not

55

a mere function of the physical environment as temperature. There are too many exceptions, for example the following.

- the reversion of the trend of a geocline, as shown above in *A. m. mellifera.*
- the two extremes in sizes are found in relatively similar latitudes: the ecotype *A. m. intermissa major* of Morocco (35°) as the largest, and *A. m. yemenitica* from Saudi Arabia (25°) as the smallest *mellifera* bee found so far.
- *A. m. lamarckii*, a small bee of tropical type is found in Egypt at a similar latitude as *A. m. i. major.*
- *A. m. ligustica* and *A. m. carnica*, occur at the same latitude but differ in size and even more in pigmentation.

These observations (and a number of others) balance the picture, pointing to the fact that recent types of bees are the result of evolution, that is of a historical process. It is rather a special case that a long sequence of gradual morphological transformations can be observed as an indication of an undisturbed evolution.

Reviewing the available data on variability in honeybees it has to be stated that the morphometric differences between races, generally considered of major significance, are mainly an external label ("for the taxonomist's convenience"). Only few of them have a proven adaptive significance (size, hair). The differences which really count, however, are of behavioral and physiological kind. Together with the external characteristics they form a compound of high individuality. Morphometrics is the most rapid and simple way to recognize this. To finish this chapter with an anecdote: Karl von Frisch, while working on his historical experiments on the "dance language" of honeybees, asked the author several times to determine morphometrically the bees he was studying in order to find an explanation for differences in behavior.

CHAPTER 5

Methods of Honeybee Taxonomy: Past and Present

5.1 The Genus Apis – A Taxonomic Problem from the Beginning

The genus *Apis* seems to be of very simple structure at first sight, an easy object for the taxonomist. Only four well-known species, and no really close relatives. However, after scrutiny, the situation appears much more complex.

Two species, *A. florea* and *A. dorsata*, are evidently "old species" with a number of highly specialized characteristics. By comparison with other taxa it was assumed that these differences are sufficient to create separate genera out of these species (*Micrapis, Megapis*; Ashmead 1904, Skorikov 1929b, Maa 1953).

On the other hand, the two species *A. mellifera* and *A. cerana* are very similar and it was not clear for a long time whether *A. cerana* (or *A. indica* at that time) should not be classified as subspecies of *A. mellifera* (Buttel-Reepen 1906, Butler 1954, Kerr and Laidlaw 1956). The same uncertainty is observed concerning the taxonomic rank and the validity of the many subspecies described.

Thus sufficient starting points for manifold confusion are provided. The four *Apis* species differ in the morphological distance and – most likely – also in age. At least two of the species, distributed over an unusually wide area with extremely different climates, show a high degree of geographic variability, ranging from slight differences in local populations to a stage of beginning speciation. The present situation is evidently that of young groups just in the process of radiation. This can be explained best by the assumption of a sudden increase in fitness within the genus and a subsequent rapid spread of a new *Apis* type over large parts of the Old World (Chap. 3).

This very brief review of the taxonomic situation of the genus *Apis* was given on the basis of our present knowledge. It took a very long time of "trial and error" to arrive at this point. In fact, honeybee taxonomy reflects the whole history of the science of systematics and its methodology in a true paradigmatic way, beginning with C. Linnaeus (1758), and arriving at the "non-Linnean taxonomy" (DuPraw 1964), which proposed substituting the conventional typological concept by abstract statistical functions.

5.2 Early History of Apistic Systematics

Honeybees are known everywhere because of their economic value and importance in human culture. They are present in most regions where man can live. No wonder, therefore, that they won early and permanent attention from the developing science of biology. New "species" were described again and again, mostly with insufficient diagnostic characterization. Since the beginning of this century the attempt was made in several reviews to amend the nomenclature (e.g., Buttel-Reepen 1906; Ashmead 1904; Skorikov 1929b; Maa 1953).

The very first discussion was caused, paradoxically, by Linné himself. In 1758 he used the name *Apis mellifera* in his basic publication "Systema naturae", Xth edition, established to include all bees known to him and also three species of wasps (Maa 1953). However, only 3 years later (1761), Linné himself changed the name "*mellifera*" to "*mellifica*" (Fauna Suecica, 2nd ed.). From this time on the scientific world had to deal with two names for one and the same honeybee: *A. mellifera*, the correct name according to the rules of priority, used in the English-speaking countries, and *A. mellifica*, the correct name according to its meaning ("*mellifera*" = honey-carrying, "*mellifica*" = honey-making), preferred in all countries of continental Europe. Buttel-Reepen (1906) argued that if Linné himself thought it appropriate to change the name, this should be respected. During the last decades, however, the name of priority is widely accepted, since the International Nomenclature Commission never agreed to a change.

The Case of "Apis Adansonii"

The multiplicity of descriptions of identical species and subspecies was almost inevitable. The early descriptions were too short and vague – mainly based only on color of dry museum specimens – to be of diagnostic use. An excellent example is the famous *A. m. adansonii* Latreille.

P. A. Latreille's description (1804) reads as follows (translation of the French original):

"6. HONEYBEE OF ADANSON. *Apis Adansonii.*

Worker bee – length of body, 0.011 m.

Description. Brown blackish, pubescent; hairs – dirty grey; scutellum, the first two rings of abdomen, the anterior half and transversal of the third – reddish, pale maroon; posterior border of the second ring, posterior half of the third, the total following rings – obscure brown.

Adanson, in whose honor I name this species, has found this insect in Senegal in trunks of trees. The individual on which my description is based being in a poor condition, I was not able to represent it" (in Plate 13)".

This description fits *any* medium-sized bee with bright pigmentation of the scutellum and the anterior abdominal tergites. Thirty years after Latreille, Lepeletier (1836) revised the genus by separating the erroneously included Meliponinae and by eliminating several dubious "species" introduced by Latreille, among others "*Apis adansonii*". Probably this "species" seemed to Lepeletier only insufficiently documented. Astonishingly, however, "*Apis mellifera adansonii* Latreille" reappeared later in the systematics of honeybees, e.g., in Buttel-Reepen's monography (1906), Friese (1909), Maa (1953), Kerr and de Portugal Araujo (1958). Now it was taken to be the "common yellow African bee", recorded from many regions of tropical Africa. Smith started morphometric studies in Tansania in 1961, and he found three clearly different varieties of bees, in different altitudes,

A.m. litorea on the coast and *A.m. monticola* in altitudes higher than 2400 m; for the most frequent variety found in East African highlands he retained the conventional name *A.m. adansonii*; but Smith's work provided the first morphometric data for this bee.

About the same time, African honeybee queens from Tansania and South Africa were shipped to Brazil under the name *"adansonii"*, the paternal stock of the now widespread "Africanized bee" of America. Thus the name given to a single poorly preserved specimen from Senegal became the generally accepted symbol of the world's most famous honeybee.

Only in the last few years, during the progress made in morphometric research into hitherto not investigated bees of the African continent, has it been shown that the honeybees of the west coast are distinctly different from those of the highlands of East and South Africa. To recognize this difference, a separate name was given to the highland bee, reactivating an existing name: *A.m. scutellata* Lepeletier 1836 (Ruttner 1975, 1981; Ruttner and Kauhausen 1985). The name *"adansonii"* should be used only for the bee of the region where it was first described: the west coast of Africa.

This story about the history of the name *adansonii* is typical for the course of taxonomic research in honeybees: deliberate assigning of names, but lack of quantitative data. There are a number of other examples, although less famous. Maa (1953), author of the most elaborate and complete monograph on the systematics of honeybees, describes the situation as follows:

"The major difficulties in re-classifying honeybees are the disappointing quality of the existing descriptions, the superabundant published names and the inaccessibility of type specimens. The last are scattered over about 17 museums and several private collections. Many of them are no longer existing or traceable. Even when being located and definitely authenticated, they are usually unavailable to students and it is often impossible to dissect out certain structures and undertake critical re-examinations.... The only possible approach to an appropriate interpretation of previous descriptions is to utilize topotypical material for the preparation of more definite redescriptions so as to prevent further confusions".

This quotation meets the point exactly. The result of 200 years of research in bee taxonomy, applying the described methods of "museum systematics", is about 600 names which were used at some time in the past for a member of the genus *Apis*, as traced by Maa (1953). Of these, 146 were regarded by him asworthy to be mentioned (but only a part of them recognized) in his list.

5.3 Attempts to Classify the Tribus Apini

The descriptive method, used a great deal by Maa himself, gave unequivocal results on the species level only with the two clearly characterized species, *A. florea* and *A. dorsata*. However, as far as the species pair *A. cerana-A. mellifera* is concerned, problems were solved in a satisfying way only in the last decade, when additionally to morphometric and behavioral, also genetic methods were applied (Ruttner and Maul 1983).

On the subspecies level a purely descriptive method based on a single or a few specimens proved to be completely inappropriate: only quantitative differences are found, and if nothing can be said about variation within colonies and populations, and between colonies and populations, no statement whatsoever can be made. However, new taxa were established with these methods; the new suspecies ("perhaps a new species") *A.m.anatoliaca* was based on the description of only three bees (Maa 1953).

On the supra-species level, *Apis* was treated as the only genus within the tribus *Apini* since Linné. Ashmead (1904) classified the same group into three genera, Megapis (= *A.dorsata* Fabr.), Apis (= *A.mellifera* L. and *A.cerana* F.) and Micrapis (= *A.florea* F.). Maa (1953), retaining this classification, split the genus *Apis* in two subgenera, Apis (= *A.mellifera* L.) and Sigmatapis (= *A.cerana* F.). At the same time he split the newly created genera and subgenera in numerous species, Apis (7 species) and Sigmatapis (11 species). Thus the formerly very modest tribus *Apini* was transformed into a notable group with three genera, two subgenera, and more than 20 species.

However, knowing at present for certain that *A.mellifera* is nothing but a species (Chap. 1), and that most likely the situation is absolutely the same with the other "genera", the number of species within the tribus is reduced again to four, or at the most five. There is no reason to complicate the classification by creating more genera than needed for grouping the existing species. Just because of the reason that categories higher than the species are to some degree arbitrary (E. Mayr 1963), it is hard to understand why a group of genera or subgenera with only one species each should be necessary.

5.4 Application of the Biological Species Concept

The best way to analyze a complex situation is to start from a clear, nondisputable point. This is, in taxonomy as well as in biology, the category of the species, the "yardstick for every classification" (E. Mayr 1963). In contrast to the majority of animals, in *Apis mellifera* the taxonomic species can be exactly matched with the biological species. By vast experience in apiculture it has been proven that even the most remote geographic varieties (e. g., from Europe and South Africa) produce fully fertile hybrids, frequently with impressive effects of heterosis (Ruttner 1983). Thus the intraspecific morphological variability is established on the safe basis of the biological species in its total range. All the characters listed as "species-specific" for *A.mellifera* are present in any of the geographic varieties of *A.mellifera.*

On the other hand, the taxon closest to *A.mellifera* in morphological and biological characteristics, *A.cerana*, was proven to be a genetically isolated species (Ruttner and Maul 1983). Therefore it can be taken for granted that *A.mellifera* and *A.cerana* are species in any kind of definition, with a great number of subspecies or geographic races (or whatever designation might be preferred).

It is true that not so much evidence is available for the other two taxa *dorsata* and *florea*, but since they occur sympatrically, each of them must have at least the rank of species. Since *within* these taxa only quantitative differences are found, of about the same range as in *A.mellifera* (Chaps. 7, 8), it can be concluded that they are also polymorphic species with a number of geographic races each.

As can be derived from these data, no problem of classification remains unsolved on the species level, except the problem of *A.(dorsata) laboriosa* with its biological and morphological pecularities (Sakagami et al. 1980; Roubik 1985). However, important data are still missing for a definitive classification of this taxon (see p. 9). The same is true also for *A. andreniformis* (p. 100).

The present state of knowledge confirms the traditional concept of the genus *Apis* with essentially four species.

5.5 Descriptive Morphometric Classification

The true taxonomic scope in honeybee systematics is the intra-specific classification. Considering the wide range of distribution and adaptations, this has the same significance as the classification of species in other organisms. As shown by earlier investigations, the conventional descriptive methods of systematics are completely insufficient at this taxonomic level, but it has taken nearly half a century since the introduction of biometrics into honeybee taxonomics for this experience to be generally recognized.

In a more advanced stage, as practised by Skorikov (1929b) and especially by Maa (1953), the previously predominant characters of color were to a great extent replaced by morphological characters. The most important ones are listed below:

1. Head
a) Relation width of head/ width of thorax
b) Position of ocelli, using the following relations (Fig. 4.6):
   postocellar line (POL) oculo-ocellar line (OOL) lateral-ocellar line (LOL)
c) Length of proboscis (scale 0.01 mm)
d) Antenna (length and relation of various segments)

2. Abdomen
General shape and structure of sternites, with description and depiction of morphological details (antecosta, apodeme, glandulus, wax plates etc.)

3. Legs
Number of bristle rows at basitarsus 3

4. Fore wing
a) Length (scale 0.1 mm)
b) Shape of radial cell
c) Cubital index

5. Hind wing
a) Length (scale 0.1 mm)
b) Rel. length of jugal and vannal lobe
c) Extension of radial vein
d) Number of hamuli

As can be seen, very few metric data (proboscis, wing) were used. All descriptions are based on individual bees, and no statistics were applied.

5.6 Start and Development of Morphometric Taxonomy

Two criteria are essential for this method, as opposed to the traditional descriptive method:

1. Means of colony characters are used as variables for statistical analysis, not the characters of individual bees.
2. Numeric data, resulting from exact measurements and analyzed with statistical methods, are used for classification.

The first morphometric study on an adequate scale with honeybees was carried out by Cochlov in 1916. This author measured the tongue length in samples of six geographic races of *A. mellifera*, taken from three colonies per race and at least 100 bees per colony (a total of 1899 bees were measured). This was the starting point of the first chapter in morphometric research in honeybees, which was written exclusively by Russian authors. Based on Cochlov's results, Michailov (1924, 1926), Alpatov (1925, 1929) and Skorikov (1929a) investigated the length of proboscis of bee colonies from many locations of the European Soviet Union. They found a gradual increase in length from north to south in the plains along a line from the Baltic Sea to the Caucasus (Chap. 4, p. 51).

The fundamental progress achieved by these authors was due to the insight into the need to apply appropriate statistical methods to analyze the geographic variability within the species and to establish a valid taxonomic system. Measurements were taken of a multitude of characters from a large number of bees out of several colonies (at least ten per taxon; Alpatov 1929), and statistical parameters were calculated. "The family is the lowest taxonomic grouping" (Alpatov 1929, p. 6), characterized not only by its averages, but also by varying correlations. At the beginning of the research program an agreement was established, with W. W. Alpatov (Zoological Museum Moscow), to take over the study of geographic variability and with A. S. Michailov (Tula Experimental Station) to investigate the influence of different environmental factors on the variation of the honeybees of a given apiary. This large-scale working program resulted in a series of papers of basic significance up to the present. Alpatov (1929) introduced a number of measurements of size in addition to the tongue: three joints of the hind leg (femur, tibia, metatarsus), length and width of the fore wing, size of the wax mirror. He recognized that the overall size of the body cannot be measured exactly, and he substituted this important characteristic by measuring single parts of the abdomen (sternites and tergites) which, of course, are closely correlated to overall size.

With this increased number of characters he found another trend in geographic variation in the plains of Russia, directed opposite to the variation of tongue length: decrease of body size from N to S. He found geographic changes in correlations, especially between length of body appendices and body size itself.

Alpatov's biometrics was initially based exclusively on characters of size. Two new sets of quantitative characters were introduced by G. Goetze (1930, 1940): Hair and wing venation. Both proved very efficient in discriminating European races.

Two characters of hair are measured: length of hair on abdominal tergite 5, and width (=longitudinal) of second tomentum (at tergite 4).

The first character in wing venation used in honeybee taxonomy was the cubital index (CI, p.72). The taxonomic significance of the relation of the section a and b of the cubital vein at the species level *(A. dorsata)* was first mentioned by Ashmead (1904), but Goetze (1930) observed the geographic variability of this character within the species *A. mellifera*. He gives the CI as quotient a: b, a value used also in our publications; Alpatov (1935 b) gives the CI as percentage (b/a.100). Later, Goetze introduced several other characters of wing venation – radial index, precubital index, "hantel index", discoidal dislocation (Fig. 4.7, Goetze 1964). The geographic variability of the latter character, the position of the distal corner of the discoidal cell in relation to the crossing point of cubital vein 3 with the radial vein, was extensively studied by Louis (1963).

5.7 Application of Morphometric Routine Methods to Actual Problems of Honeybee Biology

At least two of the characters mentioned, the cubital index and the length of hair on tergite 5, signified a decisive progress in bee morphometrics, especially as far as European races are concerned.

A.m. mellifera and *A.m. carnica* are of similar size and pigmentation, but cubital index and hair length are so different that only very little overlapping occurs even in the variation of individual bees. In consequence, both races are positioned at the opposite ends of the ramification of the "Y" (factor 2) in the structure resulting from principal component analysis of the species (Chap. 10).

The feasibility of a clear and relatively simple discrimination between the two races *mellifera* and *carnica* became very important for practical beekeeping. Various strains of the *carnica* race ("carniolans") became increasingly popular among beekeepers in Central Europe, principally on account of their gentleness, quiet temperament, and higher honey yield compared to the North and NW European *mellifera*. Thousands of queens were imported year after year to countries north of the Alps, but it proved to be impossible to maintain the strains in pure condition as long as color was used by beekeepers as the main discriminating factor. The consequence was general hybridization of bees with all the well-known undesirable side effects.

Goetze (1964) was able to develop a simple, but sufficiently accurate morphometric system to discriminate European races of honeybees for bee breeding and selection. First it was based on simple classification (low, medium, high) by estimates, but later, measurements were introduced to establish exact values of the cubital index (Fig. 6.8; Ruttner 1983).

This is one of the first and most expanded examples of using morphometric methods in selection programs, which can be applied by every interested layman. Classification of hair length is made by comparison (with help of a pocket lens) with the diameter of the wire used to fix foundations into the frame, or with the width of the first tarsal segment of worker bees, both giving the class "medium". Measurement of the cubital index can be speeded up by using a bilateral scale bent at an angle of 151° (Ruttner 1983). In this way both segments a and b can be measured simultaneously through a simple stereomicroscope (magnification

25–40 ×). Exact measurements provide the advantage of the possibility of analyzing the intracolony variation and thus to detecting even a slight, partial hybridization. This is done by simple graphic presentation (column diagram); it provides more information than calculating the standard deviation: a second peak in the lower part of a curve of *carnica* bees gives a strong indication of a partial hybridizaton with *mellifera* or hybrid drones.

This simple morphometric method provides the tool to control the results of matings and to eliminate hybrids. Thousands of nonscientific amateur beekeepers were trained in weekend classes to use this method, and the offspring of very many queens are tested every year in respect to the quality of the effectuated matings. Only since this method was introduced, has a qualified selection and breeding program been accomplished.

A better analysis of differences in wing venation was achieved by measuring certain angles instead of distances and calculating indices.

The crossing points of the veins are connected by lines, forming triangles, and the angles of these triangles are measured (DuPraw 1964, Fig. 6.9). Computing 13 angles and two characters of size (length and width of fore wing) in multivariate analysis, DuPraw was able to demonstrate a differenciation of *mellifera* worker bees in various clusters, corresponding fairly well to the geographic distribution of traditionally established races. However, DuPraw rejected the traditional hierarchical system of taxonomy by replacing it by statistical functions resulting from multivariate analysis of bee samples ("non-Linnean taxonomy"). Moreover, he ignored the intracolony variation and returned to the individual bee as the basic unit of statistical analysis.

An efficient multivariate analysis should include as many characters as possible. Moreover, the characters should be of different kind – size, hair, color, wing venation. Thus a great many of the characters used earlier and several new ones were tested in a large number of samples of different origin to establish a "standard biometry" based on 40 characters (Ruttner et al. 1978). To discriminate two or more races of a given region not all of these characters are needed, but the characters selected for discriminant analysis vary greatly from region to region. There-

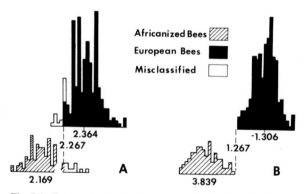

Fig. 5.1 Frequency distributions comparing Africanized and European bees. A Measurements of wax mirror width; D discriminant scores based on a set of 25 characters. The means of distributions and midpoints are indicated for each pair of graphs. (Daly and Balling 1978)

64

fore, a wide set of characters is a great advantage in studying the whole range of variation within the species *A. mellifera*. For a restricted group of races, however, the number of measurements needed for diagnosis can be substantially reduced again by stepwise discriminant analysis. This method was applied with success to a problem of high practical priority: the discrimination between Africanized and European bees in South America (Daly and Balling 1978, Fig. 5.1). The labor needed for measuring and processing data can be substantially reduced by working with a computerized device (Daly et al. 1982). For a rapid diagnosis, Rinderer et al. (1986) employed field and simple laboratory tests; Africanized and European colonies of Central America building their own comb can be identified in the field, by measuring the distance spanned by ten worker cells. By measuring a single character of worker bees (forewing length) 86% of the colonies tested (n = 136) were correctly identified, with no misidentifications. By multivariate analysis with four characters this result was only slightly improved (91% correct identifications).

The final step in establishing a reliable "natural" taxonomic classification is to include all data available on biology, ecology and genetics of the bee population under analysis ("integrated biometrics").

CHAPTER 6

Morphometric Analysis and Classification

6.1 Sampling

6.1.1 The Morphometric Bee Data Bank Oberursel

Valid classification, especially at the subspecies level, can be achieved only if a fairly good knowledge of the total variability is available. Creating taxonomic units based exclusively on local investigations may be as mis-leading as general conclusions drawn from scanty data. The main scope of this analysis was to obtain a framework – necessarily only fragmentary – of the total variability and the phenetic structure of *Apis mellifera*. To collect the samples from all the regions needed for this goal proved to be as difficult and time-consuming as analyzing them. The foundation of the collection at the Institut für Bienenkunde, Oberursel, now comprising about 1400 samples, was furnished by the valuable series of more than 400 samples from countries all over Europe and the Mediterranean gathered by Br. Adam during his many exploratory excursions in the years 1952–1976. This collection was systematically complemented and expanded in most regions of the *mellifera* area during the last 30 years by the author's activities with the help of a great number of colleagues. Besides worker bees, drone samples were also collected. Research programs on comparative behavior and reproduction stimulated a taxonomic study of *A. cerana*, based on 93 samples of this species. All data obtained from this collection are stored in the morphometric bee data bank of the Institut in Oberursel.

6.1.2 Selecting and Collecting Honeybee Samples

As shown in Chapter 4, exterior characteristics of honeybees are mainly genetically determined and highly constant, but they are nevertheless subject to certain environmental influences. Most important, bees raised early in the season and in very old brood combs with narrow cells tend to be somewhat smaller than those raised during the full season and in young combs (p. 49). Therefore, these conditions should be avoided when taking samples; sometimes, however, they have to be accepted as they are, because there is no other way to obtain them. In general, these factors do not seem to be of great influence upon our material, but a shifting in size was observed when colonies of a race with small bees were kept on European combs. The question whether the inevitable single nonconforming samples ("outliers") are due to this fact remains unanswered.

The most serious source of error concerning the authenticity of samples is hybridization by imported stock. Special care was taken to select colonies from isolated, traditional apiaries, according to the old saying "the more primitive the bee-keeper, the purer the race". In some cases samples were removed afterwards because of their known origin from a region with primarily migratory beekeeping (e. g., Ahvaz and Teheran in Iran). Since importation of queens and migratory beekeeping are an ever-increasing process in the course of modernization of the beekeeping industry, many samples of the collection are irreplaceable documents of the original bee population.

An impressive example of this source of error was given by the hard-to-obtain *A. m. sahariensis* from Morocco. We dispose of two sets of samples, one collected by Br. Adam in the original area, the oasis of the Sahara, the other from queens raised by a commercial beekeeper from original stock but mated north of the Atlas. Both sets produced separate clusters in PCA, the first closer to the Egyptian bee *lamarckii*; the second shifted towards *intermissa*. The bee taxonomist must always be alert to various sources of error. They can be minimized by careful attention and by the "effect of the great number". What counts in the evaluation of the analysis is the main cluster and the centroids, not single aberrant samples.

Principally, a sample corresponds to one colony. This requirement is best met by taking young bees (which are, moreover, in the best condition as far as hair cover is concerned). If this is not possible, as is usual when collecting from traditional hives, flight bees are taken from the entrance (most conveniently by using an aspirator). Samples supplied by entomologists rather than apiculturists are usually caught from flowers. They represent an average of the local population and not one family.

For statistical analysis the sufficient sample size calculated from the s. d., is between 10 and 20 bees. To have a safety margin we chose the standard size of 20 bees per sample. At times we had to be satisfied with a smaller number. The two segments a and b of the cubital vein (Fig. 6.8) have an unusually high s. d. and show intra-individual asymmetry which is equalized only by a larger number. For this reason this character was measured at the beginning in both fore wings, but later this was omitted to reduce labor.

Honeybees taken for a sample may be killed in any way convenient, as long as no measurement of the proboscis is to be taken. This character, however, shows higher geographic variability than any other character of size according to Bornus et al. (1976) and to results presented in this analysis (Table 6.1). To achieve a well-stretched proboscis the bees are killed either in hot water or in ether vapors. The cyanide bottle also usually works satisfactorily. With all other methods of killing, the tongue remains folded.

The samples are best preserved in a fixative liquid. Br. Adam (1983) used Pampell's fluid (acetic acid 10.0, formaldehyde 20.0, ethanol 30.0, water 100.0) and we generally took the same solution for reason of conformity. However, after decades, the acetic acid had softened the chitin and it became increasingly difficult to dissect the individuals. Therefore, Pampell's fluid was replaced by ethanol 70%. As fixative it is recommended to use either ethanol alone or in a compound without acetic acid.

Table 6.1 List of characters measured for standard morphometry

	No.	Character	Author
A. Hair			
	1.	Length of cover hair on tergite 5	Fig. 2 Goetze 1964
	2.	Width of tomentum on tergite 4	Fig. 2 Goetze 1964
	3.	Width of stripe posterior of tomentum	Fig. 2 Goetze 1964
B. Size			
	4.	Proboscis	Fig. 3 Alpatov 1929
	5.	Femur	Fig. 4 Alpatov 1929
	6.	Tibia	Fig. 4 Alpatov 1929
	7.	Metatarsus length	Fig. 4 Alpatov 1929
	8.	Metatarsus width	Fig. 4 Alpatov 1929
	9.	Tergite 3, longitudinal	Fig. 5 Alpatov 1929
	10.	Tergite 4, longitudinal	Fig. 5 Alpatov 1929
	11.	Sternite 3, longitudinal	Fig. 6 Alpatov 1929
	12.	Wax plate of sternite 3, longit.	Fig. 6 Alpatov 1929
	13.	Wax plate of sternite 3, transv.	Fig. 6 Alpatov 1929
	14.	Distance between wax plates, St. 3	Fig. 6 Ruttner et al. 1978
	15.	Sternite 6, longitudinal	Fig. 7 Ruttner et al. 1978
	16.	Sternite 6, transversal	Fig. 7 Ruttner et al. 1978
C. Fore wing			
	17.	Fore wing, long.	Fig. 8 Alpatov 1928
	18.	Fore wing, transv.	Fig. 8 Alpatov 1928
	19.	Cubital vein, distance a	Fig. 8 Goetze 1964
	20.	Cubital vein, distance b	Fig. 8 Goetze 1964
	21–31.:	11 angles of wing venation (No. 21 = A4, 22 = B4, 23 = D7, 24 = E9, 25 = G18, 26 = I10, 27 = I16, 28 = K19, 29 = L13, 30 = N23, 31 = 026)	Fig. 9 DuPraw 1964
D. Color			
	32.	Pigmentation of tergite 2	Fig. 10 Goetze 1964
	33.	Pigmentation of tergite 3	Fig. 10 Goetze 1964
	34.	Pigmentation of tergite 4	Fig. 10 Goetze 1964
	35.	Pigmentation of scutellum (Sc)	Fig. 12 Ruttner et al. 1978
	36.	Pigmentation of scutellum (B, K)	Fig. 12 Ruttner et al. 1978

6.2 Selecting Characters for Morphometric Analysis

Characters best suited should have certain qualities:

1. Distinct geographic variability as measured by the relation of intra-colony and intra-subspecies variation to the total variation within the species (Bornus et al. 1976).
2. Precision in measuring, with reasonable investment of labor.
3. Normal distribution in the population, at least when the character is used for DA.
4. Distribution across different categories of character: size of body and various appendices, wing venation, hair.

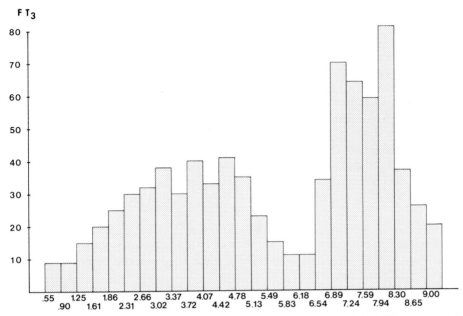

Fig. 6.1 Frequency distribution of color classes among 815 samples of *A. mellifera* (means of samples)

Compromises regarding these rules are justified if certain characters especially contribute to discrimination. All characters of hair, for example, can frequently not be measured as exactly as chitinous parts, but their inter-racial variation is extremely high and specific for certain races. Possible errors of measurement will hardly influence the result. Color is bi-modally distributed among samples (Fig. 6.1) and will most likely distort DA's if yellow and dark samples are analyzed together. Yet it is an important discriminant factor easily classified in the field, even among closely related races.

Many external structures totally lack sufficient intra-species variability; however, this has first to be found out by measuring many divergent samples. Other characters with known variability within the population, as the well-known hooks of the hind wing, are useless for taxonomic classification because no substantial variation was found between races (p. 41). However, several hundred samples had to be measured to be sure of this statement.

Luckily, a series of proven discriminatory characters was available through the work of Alpatov (1929), Goetze (1930) and DuPraw (1964). Ample experience in discriminating different races was acquired during several decades of applied morphometrics for breeding work in Central Europe (Ruttner 1983). The set of characters used in "standard morphometry", as described in this synopsis, is composed of the three authors' results mentioned above plus a few others added recently (Table 6.1).

6.3 Measurement

This program was started long before the era of the digitizer and was finished without using this certainly more efficient technique (which, of course, has no impact on results). Measurements were taken with a "profile-projector", screen 50×50 cm (Leitz) with magnification $50 \times$. Measurements of characteristics of hair and classification of pigmentation are taken under a dissecting microscope ($40 \times$). The measurement procedure is organized in such a way that all the values for each individual bee are registered in a block. One factor contributed essentially to the repeatability of the data: it was one and the same collaborator, Agnes Mohr, who performed the measurements of almost all the samples for 25 years. The quality of the data is shown by the result: two samples from the same region measured in an interval of 20 years are found close together within the same cluster.

Technique of measurement: Hair length on tergite 5 (no. 1) is measured with the undissected bee positioned in profile in front of a bright background. The general silhouette of the hair cover is measured, neglecting possibly protruding single hairs (Fig. 6.2).

Width of tomentum on tergite 4 (no. 2), relative to width of the stripe without tomentum hair (no. 3; Fig. 6.2). This measurement is taken laterally of the median line of the bee where the tomentum is broadest. A good light source has to be adjusted in a way that the short, depressed white hairs of the tomentum are clearly visible. Specimens preserved in liquid are measured submersed.

Measurements of size. All body parts needed (proboscis, hind leg, sternite 3 and 6) are stretched on a slide. If a permanent slide is wanted, the chitinous parts are imbedded in resin. If not, water-soluble material is preferable (we use gum arabic). The proboscis should be treated separately and stretched with special care.

The bent tergites are hard to flatten out. Finally we found a satisfying, simple solution: a glass rod, the thickness of a pencil, is covered with gum arabic on top, the tergites are placed across and covered with tape. In this position the pigmentation also is classified and for this reason tergites 2–4 are prepared as described. Points of measurement are shown in Figs. 6.3–6.8. Wax plates are sometimes hard to see on very light specimens. Therefore, they are stained in some laboratories with one of the stains used in microscopic technique.

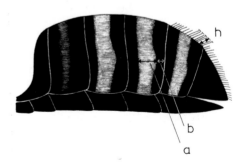

Fig. 6.2 Measurement of hair length on tergite 5 (*h*) and tomentum on tergite 4 (*a, b*)

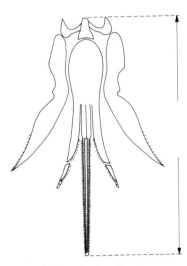

Fig. 6.3 Length of proboscis

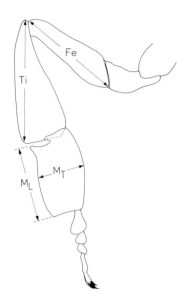

Fig. 6.4 Length of femur (Fe), tibia (Ti) and metatarsus (M_L); M_T width of metatarsus

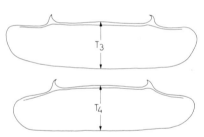

Fig. 6.5 Longitudinal diameter of tergite 3 (T_3) and 4 (T_4)

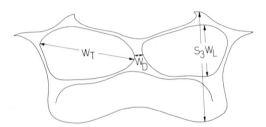

Fig. 6.6 Measurements of sternite 3: longitudinal (S_3), wax plate longitudinal (W_L), and transversal (W_T) and distance between wax plates (W_D)

During the analysis of the over-all variability of *A. mellifera* (Chap. 4) it became evident that different body parts follow different rules:

a) the trunk, increasing with latitude and altitude (Bergmann's rule);
b) appendices heavily provisioned with hemolymph and hence having a cooling effect (legs, proboscis), decreasing in length in cool climates (Allen's rule);
c) appendices with temperature protective function follow Bergmann's rule. Since in zones of similar climate allometric variations also are found, supposedly for functional reasons, it is essential to measure body parts of all three groups.

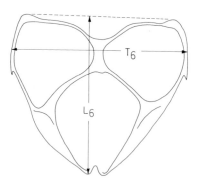

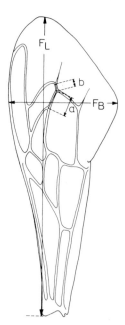

Fig. 6.7 Sternite 6 longitudinal (L_6) and transversal (T_6)

Fig. 6.8 Fore wing length (F_L) and width (F_B); distances a and b of cubital vein

When describing characters of size the use of the words "length" and "width" may sometimes appear imprecise and even confusing, because of the differing sense when used in zoology ("length" in relation to body axis) and in colloquial language ("length" in respect to the proportions of the object). This is especially evident with the narrow tergites positioned across the body axis. Therefore, whenever the meaning is not quite clear, the terms "longitudinal" and "transversal" are applied.

Wing venation. Fore wings are spread in wet condition, slightly dried and covered for measurement (length and width, distance a and b, 11 angles). For measuring angles and indices a clear definition of the measuring points is needed (which was missing in earlier papers, see Goetze and Alpatov). In our analysis crossing points of the imaginary midline of the veins were consistently taken (Figs. 6.8, 6.9). These points are marked on a transparent paper and the angles measured.

Color. For assessment of pigmentation two differing methods were proposed:

1. Measurement of the light stripes of one of the tergites (Alpatov 1929; Fresnaye 1981)
2. Classification of amount of light pigmentation according to an empirical series of pigmentation patterns (Goetze 1940, 1964; Ruttner et al. 1978).

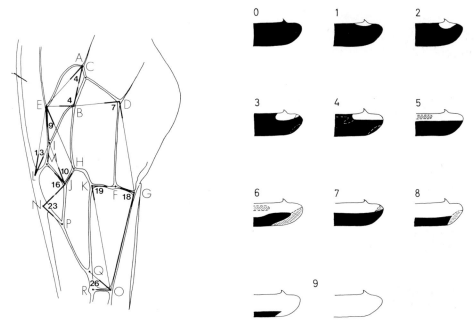

Fig. 6.9 Measurements of 11 angles (*A4 - 026*) of fore wing venation

Fig. 6.10 Classes of pigmentation of tergites 2-4

At first sight method (1) seems to be more exact and scientific. This holds true, however, only for "clearly" yellow races with uniform stripes as in Italians (Fig. 6.11, classes 7–9). As soon as no linear arrangement of pigmentation is found, but a more complicated pattern, and even more so if "shadowy" areas occur (Fig. 6.12), with method (1) the choice is either to adhere strictly to the principle by using the fixed measuring points and to make misclassifications, or to consider also the other areas, that is to work according to method (2). Moreover, "dark" populations which also show a considerable color variation, but which have light spots and no stripes (Fig. 6.11, classes 1–4) cannot be classified at all with method (1). For all these reasons, a subjective empirical classification is considered better suited to document color variability of *mellifera* races. As far as pigmentation of drones with their complicated pattern and scutellum is concerned, nothing but subjective classification seems to be possible. The distribution of pigmentation classes on tergite 2 and 3 among races is strongly bi-modal (Fig. 6.2). Closer to a normal distribution is pigmentation of tergite 4. Within races or among races of similar pigmentation a rather normal distribution is found in all tergites.

The pigmentation of the scutellum is also highly varying. At the cupola itself (Fig. 6.12, *Sc*; character 35) and the small sclerites at the base of the scutellum (metanotum; Fig. 6.12, B) and the triangular lateral segment of the mesonotum (Fig. 6.12, K; together with B, character 36) yellow areas of varying size can be found. They appear at the top or at the base of the cupola, or the whole area may

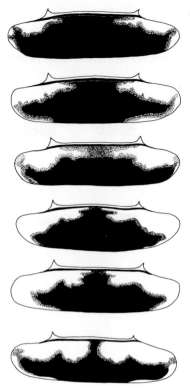

Fig. 6.11 Unevenness of pattern and intensity of pigmentation

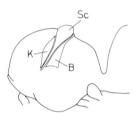

Fig. 6.12 Scutellum of honeybee. *Sc* character 35, *B, K* character 36

show various diffuse shades of brown. The classification is analogous with the tergites, from class 0 (= completely dark) to class 9 (= completely yellow). Again, not the pattern, but the overall amount of pigmentation is evaluated. The *mellifera* samples of the Oberursel collection show a bimodal distribution with one peak at class 0 and another at class 7.

Morphometric documentation of color variability seems in any case a rather approximate assessment of the natural conditions. Evidently, what is called "yellow" in Italians is not the same as it is in the Saharian and the Cyprian bee. A strictly scientific approach would be correlation of the biochemical analysis of the pigments with spectro-photometric measurements.

In drones, only 24 characters were measured: the 15 characters of fore wing as in worker bees (length, width, distance a and b, 11 angles), femur, tibia, metatarsus (length and width), and pigmentation of tergites 2–4 and scutellum. A 25th character was the number of hamuli, which was not included in the statistical analysis for reasons given on p. 41.

6.4 Statistical Analysis

As a first step, all measurements on the metric scale are transformed to a uniform scale, 1/100 mm. Then, as primary data for statistical processing, the mean of the sample with the standard deviation (s.d.) is calculated. The latter is also used as simple criterion to recognize errors of measuring and storing: each substantially deviating value is taken as an indication to check the raw data again.

It is with a good reason that means of samples and not values of individual bees are taken as operational units: the mean represents the biological unit "bee colony", individual bees are related with each other in different grades (full, half sisters), aberrant types have less weight in the analysis and the overall number is reduced in comprehensive analyses.

Sums and ratios provide a very effective tool for recognizing trends in variation and for a quick diagnosis. Various races have "key characters" which can best be represented by secondary values such as CI, index of metatarsus and slenderness, tomentum index, sum of T3+4 and of the joints of the hind leg.

Many multivariate analyses of honeybee populations have been published during the last years (Cornuet et al. 1975, 1978, 1982; Ruttner 1975, 1980a, 1981, 1986a; Daly and Balling 1978, Daly et al. 1982; Ruttner et al. 1978, 1985b, 1986; Ruttner and Kauhausen 1985; Dutton et al. 1981; Cornuet 1982a). These investigations demonstrate that clear and consistent intra-species classifications in honeybees are achieved, even within very complex groups, and also if only relatively few characters had been used (as by the Cornuet group).

Various well-known methods of multivariate analysis were applied. A two-step procedure proved to be most informative:

1. Factor analysis (especially principal component analysis, PCA) to detect separate groups (= clusters) within the population of a given area.
2. Stepwise discriminant analysis (DA) to confirm the established groups, to find the most discriminant characters and to calculate the distances between the centroids of the groups.

Finally, "ellipses of confidence" at various levels (75-95%) can be calculated according to Cornuet (1982a) to test the validity of the discrimination.

For the description of a taxonomic unit, cluster means for each unit are calculated and also ratios between certain characters (see Chap. 4 and Chaps. 11-14). A further step to achieve an improved overview of the total data is a classification of each character. In the program used for our data, the relatively few minimum and maximum values are extracted first and listed separately to reduce the range of variation. The remaining bulk is grouped into 25 classes. This listing gives an idea of the general distribution of a character (see Fig. 6.1) and shows also its geographic distribution.

Not only the morphometric values by themselves but also varying correlations among them may be specific charateristics. The existence of allometric variability was already shown by Alpatov (1929, Fig. 4.1). The overall correlations for 11 characters of different categories, extracted from a factor analysis of 245 *mellifera* samples, are shown in Table 6.2. Highest are correlations of size measure-

Table 6.2 Correlations between 11 characteristics

No. of character	1	4	6	7	10	16	17	19	23	27
Character	Hair	Prob	Tib	TarsL	T 4	St6T	FWL	Veina	D 7	I 16
4. Proboscis										
6. Tibia	36	82								
7. Metatars L	24	76	89							
10. Tergite 4	32	75	88	80						
16. Stern 6 tr	66	76	84	78	84					
17. Fore wing L	60	76	89	80	83	85				
19. Cub. vein a	15	33	30	36	32	29	48			
23. Angle D7	13	12	06	−12	−10	04	−35	34		
27. Angle I16	−03	−25	−04	17	−04	17	03	03	−27	
32. Color T2	−28	−30	−38	−44	−41	−47	−50	−10	−03	−18

ments varying between 80 and 90. Generally, higher correlations were found when comparing longitudinal with longitudinal rather than longitudinal with transversal measurements.

Hair length is correlated at varying levels with size, indicating a link between Rensch' rule and Bergmann's rule (longer hair in cooler climate; Chap. 4). Length of vein (a) is somewhat correlated with size, highest with wing length, naturally. Wing venation angles do not show a general trend of correlation with other characters. Color is negatively correlated with size, also slightly with hair length, conforming the observation that northern and mountain types tend to be large and dark.

Standard programs used:

BMDP Biomedical Computer Programs P-Series (1979). (Ed.: Dixon W.J.; Brown M.B. University of California Press)
SPSS Statistical Package for the Social Sciences (2nd ed. 1978). (Ed.: Nie N.H.; Jenkins J.G.; Steinbrenner K; Bent D.H.). Mc Graw Hill, New York

6.5 Classification

"Morphometric methods are powerful research tools when used in the context of sound biological knowledge" (Daly 1985). This sentence, with special emphasis on the second part, is the right introduction to the following paragraph. Classification at the subspecies level necessarily includes strong arbitrary and subjective elements – quite in contrast to the species level for which at last a clear and generally accepted definition has been found.

The arbitrary status of the taxon "subspecies" was very obvious in the period of descriptive taxonomy and its frequently imprecise diagnosis. Statistical methods allowing calculation of discriminatory functions and also their confidence seem to shift microtaxonomy to another scientific level with the attribute of increased exactness. But does this hold true also from the standpoint of taxonomic classification?

Statistically, subspecies can be discriminated with a very high significance. Populations within one and the same subspecies can also be morphometrically discerned, even if they are found to be only 8 km apart (Cornuet et al. 1978); yet single colonies can be discriminated from others, not only by the averages of their characters but also by the coefficients of correlation of one body part to the other. "The family is the lowest taxonomic grouping" (Alpatov 1929). The elegance and power of statistic tools permit classifications below the level of the genetic population and include, therefore, the possibility of taking a dominant position in the process of infra-species classification, similar to that of the morphological "type" in a former definition of the species.

When many phenetic "types" can be separated within original populations of honeybees by sophisticated morphometric methods, what level should be taken as criterion for the taxon "subspecies"? "Whenever a thorough biometric-morphological analysis established a mean difference between samples, this was considered sufficient justification to describe a new subspecies" (E. Mayr 1963). This approach might (and did) lead to a "wildgoose chase" for new subspecies, seriously impairing the usefulness of the subspecies category. In a biologically meaningful classification the morphometrical analysis has to be supplemented by essential attributes of a biological unit like, e.g., specific ecological adaptations, behavioral characteristics, and clear geographic demarcation. It is this complex view of morphological and biological characteristics which reduces biased positions when considering intra-species classification.

For the biologist the subspecies is primarily a grouping of special adaptations in physiology and behavior to a certain type of environment with secondary variations of the external characteristics. Apicultural researchers are in a better position than others since they know from long-standing personal experience that correctly defined geographic races have biological individuality and are not mere arbitrary creations. The true problems, however, are found in groups of neighboring "races" with morphometric similarity, and in the variability within the subspecies.

Subspecies are not a uniform group of individuals but are composed of a varying number of populations. Since processes of evolution take place within these populations, a considerable degree of variation also exists. Therefore, not every single individual (= colony) can be assigned to the subspecies morphometrically, yet the population as a whole can. Since transition to other races can be expected for various reasons also, traditionally at least 75% of the individuals of a population should show the attributes of the subspecies to which it is assigned (Huxley 1939). Populations within a subspecies with a large area of distribution comprising various climatic zones are likely to show different ecological adaptations and also morphometric features. The crucial question is: Should these subunits be recognized as taxonomic units, as a member of a "circle of races" ("Rassenkreis", Rensch 1926)?

An example may illustrate the problem. *A. m. mellifera* is one of the best investigated subspecies of the Western honeybee. Every specialist will recognize the major morphometric and behavioral characters at first sight, whether the bee comes from the Provence, from Scandinavia, or from the Ural. Yet in France several local populations were found to be differing in physiology and morphometrics (p. 239). It can be assumed that many more varying populations could be found in other parts of the subspecies area. Should these populations be classified as geographic races or not?

It would be extremely hard, however, to delineate the distribution of these populations, to agree on the acceptable level of overlapping and remain informed about the presumably large number of other local or ecological populations. The taxonomic rank of these units is certainly distinctly lower (evaluated by the morphometric and behavioral distance) than that of the geographic race "*mellifera*". A similar situation is observed in other subspecies with wide distribution such as e. g. *A.m.carnica* (p.251), *meda* (p.191), *yemenitica* (p.203) and *A.cerana indica* (p.157).

Keeping in mind the generally recognized attributes of a subspecies (=geographic race) and established also for the "traditional" races of honeybees, it seems justified to apply these principles consistently in honeybee taxonomy. Again, this insect provides an excellent starting point: quite a number of races are well investigated and known by apicultural experience (*ligustica, carnica, mellifera, caucasica, intermissa*). Therefore, a standard for the rank of a subspecies in honeybees was available for the taxonomic classification of less known groups.

Major factors in morphometric evaluation were distance of centroids, amount of overlapping and number of samples (at least ten; Alpatov 1929). Further criteria were size and delimitation of area of distribution and ecological and behavioral characteristics. The result is a preliminary list of 24 subspecies in *A.mellifera* and of four in *A.cerana*. One subspecies, *A.m.armeniaca*, is incorporated only reluctantly because of the minor statistical distance to *A.m.anatoliaca* and the small number of samples. A.m.major, introduced by the author in 1975 because of its surprisingly divergent morphometrics, was withdrawn as subspecies due to lack of information on biology and to the assumed restricted area, and is mentioned only as "microgeographic race" within the area of *A.m.intermissa* (E.Mayr 1963). A.m.taurica (Alpatov 1938b) is treated similarly, in as much as only one single publication exists about it (50 years old). The present list has to be taken as a combined result of all information available to date and certainly will require future adjustments. This is especially true for *A.cerana*, represented by only 93 samples, although the fact has to be stressed that the island populations of S Asia were isolated much later (post-Pleistocene) than the Mediterranean races of *A.mellifera*. Local populations with specific characteristics are mentioned in the section on the race to which it belongs. They may be named by the region of origin, in addition to the subspecies name, but not with a separate taxonomic name.

Of special interest is the importance of ecological factors in isolating geographic races of honeybees. *A.m.monticola* from Kilimanjaro could well be called an "eco-race" would it not be for the considerable morphometric and behavioral differences. It can be demonstrated that ecological adaptations exist in all geographic races and local populations.

Ecological diversification, therefore, has to be considered as an essential part in taxonomic intra-species radiation, instead of being a separate category. "It is not permissible to make a distinction between geographical and ecological barriers" (E.Mayr 1963).

CHAPTER 7

Apis florea Fabricius 1787:305

7.1 Introduction

The Dwarf Honeybee has found its ecological niche in the stratum of dense bushes and small trees of the tropics. This biotope requires small nests and small body size, and special adaptations for survival. *A. florea* is the most peculiar *Apis* species in several ways (Free 1981). Generally it is regarded as a species on a relatively low phylogenetic level within the genus. In several characteristics, however, it is certainly a very differentiated species, remote from the general line of evolution observed in the other *Apis* species.

In the periphery of the species range, the ability of *A. florea* to survive in a very hot and dry climate is one of its most outstanding traits. On the Persian Gulf it tolerates summer temperature maxima of 50°C and more (Whitcombe 1984a; Ruttner et al. 1985a). No other honeybee occurs in this region. In this same area interesting behavioristic adaptations are observed to overcome cold periods during winter in spite of the habit of nesting in the open with one single comb.

7.2 Area of Distribution

A. florea is the honeybee of the lowlands of South Asia, in some regions the only one. It extends much further to the West than the other Asiatic *Apis* species (Iran, Oman, probably also Iraq and Abu Dhabi, Whitcombe 1984a), but not as far to the North and East (Fig. 7.1). The main habitat of the species is Pakistan, India, Sri

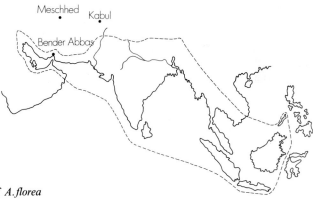

Fig. 7.1 Distribution map of *A. florea*

Lanka, Thailand, Indochina, Malaysia, parts of Indonesia (Sumatra, Java, Borneo), and Palawan (but not the other islands of the Philippines) in altitudes up to 500 m (Maa 1953; Free 1981). On the Persian Gulf coast the occurrence of *A.florea* coincides exactly with the strip of subtropical climate (less than 5 days of temperature below $+5°$ C). During summer, colonies were observed up to 900 m in Iran (Ruttner et al. 1985 a), up to 1900 m in Oman (Whitcombe 1984 a), India (Muttoo 1956) and Thailand (H. Pechhacker, pers. commun. 1985). *A.florea* lives sympatric with *A.dorsata* and *A.cerana* in the major part of their areas, but with *A.mellifera* only in NE Oman. The *mellifera* bee of this country, *A.m.yemenitica* (a race of African origin), however, was allegedly imported only 300 years ago (Dutton et al. 1981).

7.3 Morphology

7.3.1 General

No overlapping occurs in any of the measurements of size with the next largest, *A.cerana* (Fig. 4.2). As far as the functional body parts are concerned, *A.florea* could be regarded as a minute copy of this species. The relations between the

Fig. 7.2 *A.florea* queen (from Iran) with court of worker bees on top of the honeycomb. (Phot. U. Eidam)

Fig. 7.3 *Florea* drones among worker bees (from Pakistan)

Table 7.1 Diameter of worker and drone brood cells (w, d; in mm)
(Data from Zander, Weiss 1964; Ruttner et al. 1985 a)

	florea		*mellifera*	
	w	d	w	d
	2.9	4.6	5.3	6.9
Ratio	1:	1.59	1:	1.30

three Asiatic species *florea – cerana – dorsata* in body size are 1.0:1.33:2.39, and in fore wing length 1.0:1.27:2.09 (Ruttner 1986 b). Nonfunctional characteristics show also distinct differences. Out of 11 angles of wing venation measured during our morphometric analysis of 12 *florea* samples with 20 workers each, 9 differ significantly from *A. mellifera* and 5 from *A. cerana*. Another significant difference is the very narrow distance between the two wax plates of a sternite.

A very conspicuous characterisic of *A. florea* is the striking difference in size between worker bees and sexuals (Figs. 7.2, 7.3). This inequality is much smaller in the other species (p. 8). Differences in body size are reflected also in the diameter of brood cells (Table 7.1)

The small size of the worker caste can be understood as adaptation to a special biotope, especially to the nesting site and foraging. The sexuals might have retained the original "*apis*" size for the sake of their function in the social community.

The most obvious characteristic except size is the basitarsus 3 of *florea* drones, which is transformed to a mitten-like organ (Fig. 7.4). The "thumb", separated from the main part by an incision, is an empty chitinous case without any muscles. The inner side of the "thumb" is covered by strong, sharp spines, which correspond to a cover of soft, curled feather hairs on the opposite surface of the mainstem of the basitarsus (Fig. 7.5; Ruttner 1975b).

This structure was called "clasper organ" because it fits exactly on the queen's hind tibia (Fig. 7.6). A heavy brush on the basitarsus of *dorsata* drones seems to have the same function (p. 108). However, the actual position of *florea* queens and drones during mating has not yet been observed. The hind leg of a *florea* drone was first depicted by Horne and Smith (1870) and described by Drory in 1888 (Buttel-Reepen 1903).

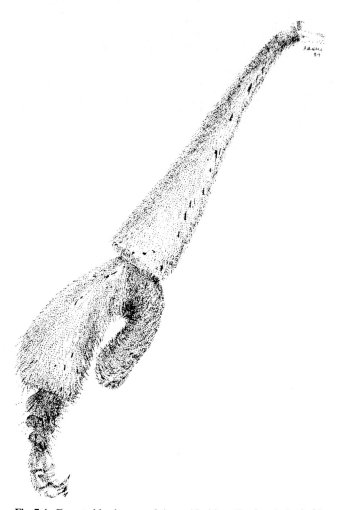

Fig. 7.4 Furcated basitarsus of drone hind leg. (Design A. Aarhuis)

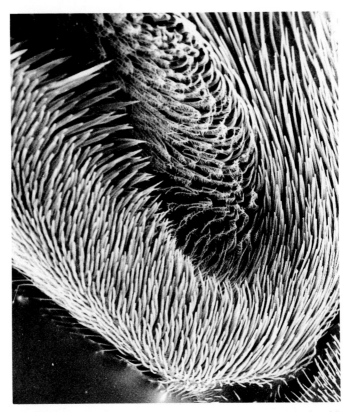

Fig. 7.5 Clasper organ of the drone: stiff bristles on the inner surface of the basitarsal "thumb" and smooth plumose hairs opposite (Ruttner 1975b)

7.3.2 Genitalia

a) Queen (Fig. 7.2). The diameter of the spermatheca (with cover) was measured in three queens from Sri Lanka; it was found to be 0.83 mm in two cases and 0.96 mm in one case – only somewhat smaller than in *A. mellifera*. The number of spermatozoa in the spermatheca was found to be 380,000 in one queen shortly after she had been taken from her nest in Sri Lanka. Another queen of the same origin maintained a colony in a flightroom of the Institute in Oberursel for a whole year (August 1972–August 1973). During experiments the colony lost many bees and the queen stopped egg laying. When she was dissected, 922,000 spermatozoa were counted.

b) Drone (Fig. 7.3). Since no illustration exists of the male genitalia except one of the inverted endophallus (Bährmann 1961a), these organs will be described in some detail.

The genital glands in general are very similar to those of *A. mellifera*. The bean-shaped testes of immature drones, measuring 1.25×0.70 mm (Fig. 7.7) are situated

Fig. 7.6 Supposed function of the clasper during mating. (Design A. Aarhuis)

in the frontal part of the abdomen, with their frontal ends corresponding approximately to antecosta 3. The mucous glands are much smaller than those of *A. mellifera*, even when compared with abdominal size. The number of spermatozoa produced by one drone is not known.

The everted endophallus (Fig. 7.8) looks very different from that of the other species, though all the single structures, including the hairfields, can be identified. There is only one pair of long, curved cornua, thick at their base and narrowing towards the end; as in the other species, they show a cover of a sticky, orange-colored substance. The fimbriate lobe has a clumsy, cone-like shape. The end of the endophallus is not inflated to a bulb, but progressively reduced to a narrow, curved tube (Fig. 7.9). The total length of the everted endophallus, from its base to the curved tip is 6.25 mm.

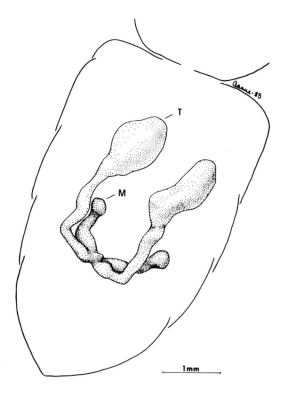

Fig. 7.7 Genital glands of *florea* drone.
M mucous gland, *T* testis

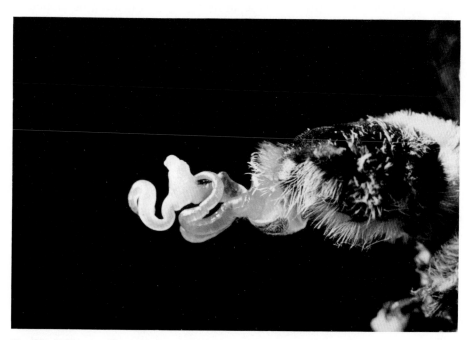

Fig. 7.8 Fully everted endophallus of *A. florea*

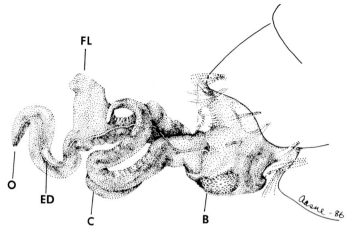

Fig. 7.9 Details of endophallus. *B* bursal hair plaque, *C* bursal cornu, *ED* ejaculatory duct, *FL* fimbriate lobe, *O* orifice of duct (for homologous structures see *A. cerana*, Fig. 9.11). (Design A. Aarhuis)

7.4 Biology and Behavior

7.4.1 Physiology of the Individual Bee

Knowledge in this field is poor, although the dwarf bee is an ideal subject for experimentation, in the open as well as in flight rooms. The mean development period of workers and drones is very close to that in *A. mellifera* (Table 7.2).

The size of the egg is much the same as in *A. cerana indica* (0.416 × 1.675 mm, resp. 0.403 × 1.750 mm), but it is smaller than in *A. dorsata* (Woyke 1975) As to the life span of the individual worker bee, only limited observations of a colony kept in a flight room at the Institute in Oberursel are available (observer B. Bendig). The average life span of ten marked bees out of a colony with brood was 61.2 days. This is about 2.5 times as much as in *A. mellifera.* The last bee died 118 days after marking. This surprisingly long life span of *florea* workers was confirmed through all the years when this species was kept in flight rooms in Oberursel. If the same is true also for feral life, this characteristic would contribute a great deal to this bee's faculty of survival under adverse conditions.

The individual temperature tolerance seems to be higher compared to *mellifera,* as concluded from the daily flight activity: In the morning, *florea* field bees start about 2 h later (at ambient temperature of 18°C vs. 10°C in *mellifera*), but

Table 7.2 Duration of development of *A. florea* (in days). (Sandhu and Singh 1966)

	Egg	Larvae	Pupae	Total
Worker	3.0	6.4	11.2	20.6
Drone	3.0	6.7	12.8	22.5

86

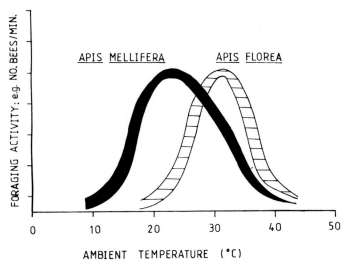

Fig. 7.10 Diagram of daily flight activity of *A. florea* and *A. mellifera.* (Whitcombe 1984a)

continue also during the hours with temperatures higher than 40°, when *mellifera* stops all flight activities. Therefore, a significantly different daily flight pattern is the result (Fig. 7.10; Whitcombe 1984a).

7.4.2 Nest and Nesting Site

The special structure of the *florea* comb, being basically different from the comb structure of the other species, is conditioned by the substratum of the comb. The three other species fix their comb onto a broad, horizontal surface. They start with the vertical mid-rib, which is extended laterally with hexagonal cells. The first row of cells helps to fix the comb to the substratum, adhering to it with two of its lateral walls (adhesive cells; Fig. 7.11).

The most frequent nesting site of *A. florea* is the thin branch of a bush or a small tree. To attach the comb to this narrow support in the way described before would be impossible. Thus the *florea* comb encircles the branch completely (Fig. 7.12). Construction is started with the hexagonal cell pattern and placed immediately on the surface all around the branch. From this basis the cells are extended more or less horizontally to both sides. Laterally they become very deep, 20–30 mm (Thakar and Tonapi 1962; Ruttner et al. 1985a); on top the cells are reduced in depth, thus creating a more or less flat surface. Where the opposite cells meet, above and below the branch, a common bottom is formed, separating the cells and becoming the "foundation" of the comb. in this case not as the primary element of construction, but as secondary effect (Fig. 7.13).

The cylindrical cell structure round the branch, succesively filled with honey, provides the "dance floor" on its flattened top ("Tanzboden", Lindauer 1957) for the communicating field bees. Beneath the branch, the comb is abruptly reduced

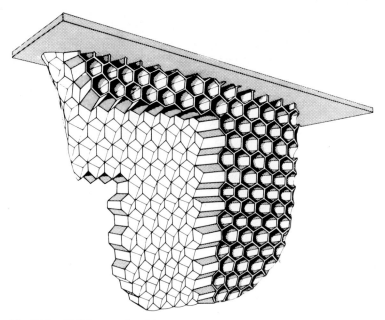

Fig. 7.11 *Mellifera* comb with cells constructed on both sides of primary midrib. (Design M. Kreuder)

in diameter (16 mm). This vertical part of the nest with very regular, small cells (2.7–3.1 mm) is the brood comb. The storage capacity of the honey cap or crest is 500 g up to 1000 g or more of honey (Muttoo 1956; Tirgari 1971).

The size of the *florea* comb varies greatly. Even if only "mature" combs are compared (combs with queen cells, indicating that the colony has stopped to increase), comb size varied according to Ruttner et al. (1985a) in South Iran between 130 × 130 mm and 270 × 360 mm (= 265–1527 cm²). Whitcombe (1984a) observed an extremely large comb (350 × 740 mm). Combs with 600 cm² of brood were reported. This corresponds to more than 14,000 occupied brood cells on both sides of the comb. The much larger drone cells (0 4.2–4.8 mm) are generally found in the lower part of the comb, the queen cells at the bottom (Fig. 7.13). Little is known about the properties of *florea* wax: the spectra obtained from a gas-liquid chromatography are different in number and height of peaks from that of other species (Tulloch 1980).

Corresponding to the decreasing body size from north to south (p. 101), the diameter of worker brood cells shows a geographic variability in the same sense. The number of cells counted per dm² was 1190 in northern India, 1240 in central and 1560 in southern India (Muttoo 1956).

A. florea also nests in small caves with wide apertures or in sheltered areas of buildings, or even in wells far below ground surface. Here the comb support can be any kind of niche in the wall, eaves, and boxes, etc. In treeless regions in Oman they also use rock cliffs as nesting sites. In these situations the attachment of the comb to the support differs from that to a branch, the construction type, however,

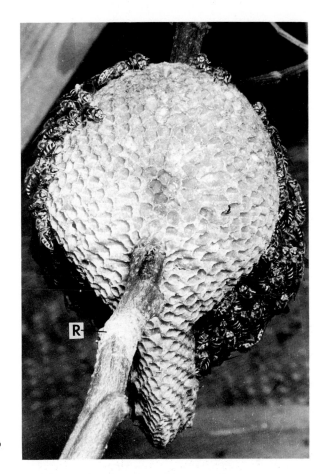

Fig. 7.12 Nest of *A. florea* with honey comb ("dance floor" on top) and brood comb. *R* Ring of resin. (Photo M. Lindauer)

remaining unchanged. In any case, a hexagonal cell pattern is started on the support (Figs. 7.14, 7.15). If on a vertical wall, the comb is attached laterally to it, the crest thus becoming unilateral (Fig. 7.14). If on a horizontal ceiling, a vertical comb is brought about by gradually shifting the axis of adjacent cells in opposite directions (Fig. 7.16). In this case the honey cap is positioned on the side of the brood comb immediately attached to the ceiling.

7.4.3 Defense Behavior

The general position of *florea* nests relatively close to the ground caused the evolution of defense strategies specifically against small predators, mainly ants and other insects. An obviously effective, sophisticated device to protect the nest are two rings of "insect lime" round the branch, one on each side of the comb (Figs. 7.12, 7.13). This glue, a kind of propolis but more sticky than that used inside the hive of *A. mellifera*, is found at *florea* nests throughout its area. Worker

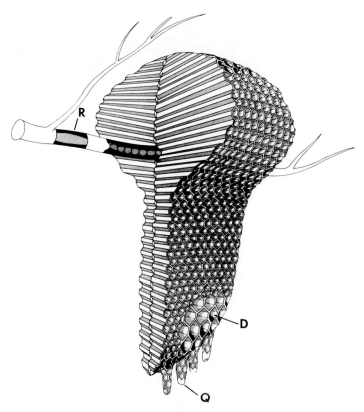

Fig. 7.13 *Florea* comb with cylindrical honey comb round a branch and flat brood comb beneath. Secondary midrib where opposite cells meet on top and below a branch. *D* Drone comb, *Q* Queen cell, *R* Ring of resin. (Design M. Kreuder)

bees were observed to take care of these bands regularly by roughening and keeping them sticky, perhaps by adding some secretion (Lindauer 1956, 1957). The glue has a repellent effect on ants, but frequently they stick to it (Fig. 7.17).

The disturbance of the bees at the comb elicits a "hissing behavior" very similar to that described with *A. cerana* (p. 134). If an intruder (e.g., a foreign bee) lands on the comb, it is attacked by the worker bees and sometimes clumped (Free and Williams 1979). *A. florea* produces isopentyl acetate as alarm pheromone, but not 2-heptanone. It shares, however, only with *A. dorsata* a third alerting pheromone with low vapor pressure, 2-decen-l-yl-acetate, suited to mark a spotted predator (Veith et al. 1978).

Attacking an intruder on the wing and stinging is in general not easily provoked. Usually the Dwarf Bee is described as "mild" and the common technique to collect honey consists in brushing the bees with the bare hand off the comb. Dutton and Free (1979) and Whitcombe (1984 a, b) illustrate in their publications how *A. florea* is manipulated by local beekeepers in Oman without any protective clothing. It is in accordance with this low level of defense reaction that Morse et al. (1969) found in *A. florea* the lowest amount of iso-amylacetate (0.1–0.5 mg) of all *Apis* species.

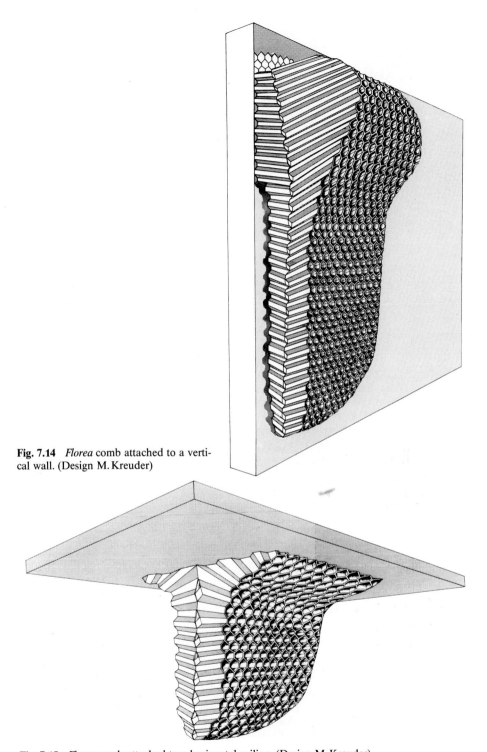

Fig. 7.14 *Florea* comb attached to a vertical wall. (Design M. Kreuder)

Fig. 7.15 *Florea* comb attached to a horizontal ceiling. (Design M. Kreuder)

91

Fig. 7.16 *A. florea* nest, 2 days old, attached to the top of an open wooden box. Cells with stores (pollen and honey) laterally to the brood comb (with eggs). (Photo R. Whitcombe)

Fig. 7.17 Ants stuck to the ring of resin close to a *florea* nest. (Photo T. D. Seeley)

Fig. 7.18 *Florea* nests, with worker and drone brood, for sale on the market of Bangkok. (Photo G. Schneider)

If abruptly disturbed, usually a colony will rather abscond than defend, although sometimes it may attack fiercely (Morse and Benton 1967). Absconding, too, is a way of defense to permit survival if the nest is close to its main predator, man. Since in most cases the whole nest is taken together with the brood in order to obtain the appreciated *florea*-honey (Fig. 7.18), the readiness to abscond and immediately start a new nest elsewhere gives the best chance of keeping losses to an acceptable level (Fig. 7.19). The ability of the queen to take to the wing any time, even while in oviposition, fits this behavior. The predation toll may be extremely high in some regions; in Thailand Seeley et al. (1982) found during a 7-month observation period that 25% of the *florea* nests were destroyed or abandoned every month. The maintenance of the population in spite of this amount of predation can be explained only by frequent survival of the colony and a high re-

Fig. 7.19 *Florea* workers, dismantling an abandoned comb and transporting wax particles in the pollen basket to the new site. (Photo T. D. Seeley)

production rate. In line with a primarily passive defense strategy lies also the preference for hidden, protected nesting sites (Whitcombe 1984b). If the sheltering leaves are removed, artificially or during the dry season, the colony will abscond (Seeley et al. 1982).

When the nest is disturbed, drones appear on the surface on top of the crest and take off (Fig. 7.3). Virgin queens were observed to behave in the same way.

7.4.4 Regulation of Microclimate

The insulation of the single comb, mostly in the open, is effected by a thick curtain of bees on both sides of the brood comb, variable both in thickness and in distance from the comb, depending on ambient temperature (Lindauer 1957; Whitcombe 1984a). At ambient temperatures between 18 and 32.5°, the central area of the brood is maintained between 33 and 38° (Free and Williams 1979). Thus the capacity of thermoregulation of *A. florea* is inferior to that of *A. mellifera*, as demonstrated convincingly also by Whitcombe (1984a). In the coastal plain of Oman (Khabura), *A. florea* maintained a steady temperature just above 34° during day time (ambient temperature 32–36°). In the early morning (ambient t 21°) the temperature of the brood area dropped to 32°, that of the honey comb to 25°. In the mountain village of Saiq (alt. 1950m), however, with an ambient temperature oscillating between 14° and 20° during day time, the temperature of the comb surface (no brood present) of *A. florea* fluctuated between 23°C and 31°C, thus dropping

94

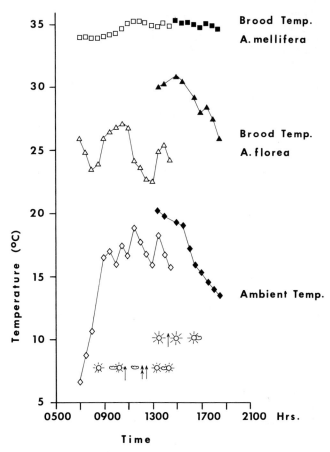

Fig. 7.20 Cluster temperature of a broodless *florea* colony and brood temperature of a *mellifera* colony at Sayq (alt. 1950 m, Oman). (After Whitcombe 1984a)

below the physiological limits of brood rearing (difference to ambient t 6 and 13°; Fig. 7.20). At the same time the temperature in a *mellifera* colony with brood remained constant between 34–35° C.

The incapacity to maintain a constant high temperature level during periods of low ambient temperature explains much about some characteristics of the area of distribution: restriction to tropical lowlands with rather balanced thermic conditions. The down regulation of high ambient temperatures to brood temperature level by fanning, variation in density and position of the bee curtain and evaporating (Lindauer 1956, 1957; Acratanacul 1977; Whitcombe 1984a) seem to be far more efficient compared to heating, as proved by the presence of this bee on the coast of the Persian Gulf with maximum temperatures higher than 50° C, where no other honeybee can exist (Ruttner et al 1985a). Lindauer (1956, 1957) measured a brood comb temperature of 34–36° C while the colony was in full sunshine with a shade temperature of 42° C.

This reduction of temperature is evidently the combined effect of a number of factors: reflection of radiation by irridescent wings; isolation of the comb by the tightly packed bee curtain, leaving an air pocket inside; spreading water on the comb, and fanning. An important factor for maintaining the microclimate of the nest is the amount of shading, which was found to average 75% in open-air nests in Oman. This average of shading seems to be a compromise between needs of microclimate, protection against predators, and orientation; the colony is believed to pursue an equilibrium by eventually changing its site. The SE sector of a tree is clearly preferred as nesting site, offering higher temperature in the morning and shade in the afternoon (Whitcombe 1984a).

Seasonal migration is another means of creating a compatible micro-climate. In Iran, at the western border of the *florea* area, the colonies regularly change their summer residence in the shade of a tree to a sunny sheltered place at a house wall. The exposition of the comb is turned from N-S (summer nesting site) to E-W (Tirgari 1971). The advantage of this behavior is evidently so great that it makes up for high losses due to predation by man. It could well be that sheltering by human construction and artificial water supply were the prerequisites for the colonization of the Persian Gulf coasts by *A. florea*.

Frequent short-distance migration is enhanced by a special behavior trait of *A. florea*: these bees take along all the stores and even the wax of the combs (Fig. 7.19). K. v. Frisch once wrote to the author: "It is fascinating to think that bees take along their furniture when they move". *A. florea* shares this behavior of transporting nesting material from the old site to the new nest with Meliponae (in this case during the process of swarming; A. Wille 1983). Absconding may occur at any time of the season, with a peak, however, in a dry and hot period with poor supply of food and water. The residence of a *florea* colony on a certain site, therefore, is narrowly limited. Mean residence time in Thailand was 2 months in the dry season and 5 months in the wet season (Seeley et al. 1982). In Oman, Whitcombe (1984a) observed a mean residence time of 101 days with his experimental colonies, one of them staying at the same site for more than 1 year.

Certain observations indicate that also long-distance seasonal migrations occur in *A. florea* (in summer from the hot plain to the hills 800–1500 m high; Muttoo 1956; Ruttner et al. 1985a).

Another problem for free-nesting colonies is the heavy monsoon rains. *A. florea* relies to a large extent on self-protection: a tightly packed layer of parallel-oriented bees on the surface and the temporarily waterproof haircover and the waxed body surface of the individual bee. However, long-lasting rains may break this protection, at least for individual bees: Lindauer (1956, 1957) observed soaked bees lying dead on the top of the comb.

mellifera   florea

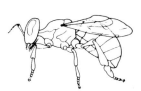 

Fig. 7.21 Body posture of *A. mellifera* and *A. florea* during waggle dance (Gould et al. 1985)

7.4.5 Field Activities and Communication

A. florea bees are the most important pollinators of field crops in the plains of India and Pakistan. Atwal (1970) observed 73–84% *florea* bees among the insects visiting *Brassica campestris* var. Toria and Sarson, *Brassica juncea* var. Raya, and *Medicago sativa*. Sihag (1982) reached the same conclusions, since *A. florea* is counted among the most efficient pollinators of agricultural Umbelliferae, as well as of *Allium* and *Brassica* sp. *Florea* foragers visit a broader array of plant species (Koeniger and Vorwohl 1979; Whitcombe 1984a) than larger species, thus making up for the short flight range, which usually does not exceed 500 m (Lindauer 1957; Tirgari 1971).

The communication among flight bees has been well known since the classical study of Lindauer (1956, 1957): *A. florea* performs round and waggle dances on the horizontal plane of the honey crest and is able to communicate direction of and distance from a honey source, but it is not able to transpose the indication of direction to the vertical plane. Therefore, in order to perform oriented dances, the bee has to see the sky. The tempo of the runs is slower than in the other *Apis* species for the same distance. During the horizontal waggling motion, *florea* workers curve the abdomen upwards (Fig. 7.21), with a superimposed dorso-ventral motion. This exaggerated body posture (observed also in the other open-air-nesting species, *A. dorsata*) is supposed to facilitate visual perception by recruits (Gould et al. 1985).

In the last few years, however, *florea* combs fixed to a ceiling (Fig. 7.16) and dances on a vertical comb have been observed (Free and Williams 1979; Gould et al. 1985). Even oriented wagtail dances by bees which were not able to see the sky have been seen (Koeniger et al. 1982). In this situation bees use learned land marks of close surroundings, which have to be correlated with the direction of the food and the sun's azimuth course, for orientation ("canopy orientation"; Dyer 1985a). With this flexible system of orientation it now seems inappropriate to regard *A. florea*'s dance merely as a "primitive" version of the honeybee dance language. Nevertheless, no transposition of the horizontal direction into gravity orientation in the vertical plane was observed in *A. florea*. This fits into the observation of Jander and Jander (1970) that the geotactic behavior of this species is similar to that of *Bombus* and *Trigona* ("progeotaxis"), while the other three *Apis* species show a more advanced behavior ("metageotaxis"). The progeotaxis of *A. florea*, however, is of a special kind, since gravity is registered by the petiolus organ (as in *A.mel-lifera*) and not by pedal receptors as is the case in *Bombus* (Horn 1975). Instead of interpreting the special reaction of *A.florea* to gravity as "ancestral", it could as well be explained as a derived simplification of the dance behavior of the other *Apis* species (N. Koeniger 1976b).

7.4.6 Interspecific Relations

A.florea competes well on a common feeding place with other *Apis* species in spite (or on account?) of its inferior size. The Little Bee takes its stand also in the process of interspecific robbing (Koeniger and Vorwohl 1979). The thick curtain of "idle" bees round the comb not only provides the microclimate needed for the brood, but also protects against small aggressors (insects of any size) by a simultaneous mass reaction (Koeniger and Fuchs 1975). Even the recent coexistence with the allopatric *A.mellifera* does not seem to create difficulties.

7.5 Reproduction

The daily egg rate of a queen during high season of colony growth was reported to be 350. From a brood area of 600 cm^2, as found by Kshisagar et al. (1980), it is evident, however, that this figure can sometimes well be doubled. Inconsistent data were published about the number of bees present in an average *florea* colony. See-ley et al. (1982), however, reported precise figures about 12 colonies from Thailand. They found an average of 6271 workers with a high standard deviation of 4927. In Oman, Whitcombe (1984a) counted almost the same number in one colony. The number of "up to 30,000" given by Tirgari (1971), therefore, has to be regarded as a rare maximum.

The development of the *A.florea* colony is solely directed towards the swarming process, which frequently initiates the complete dissolution of the colony. This trend does not appear so clearly in the other *Apis* species. The bees are constructing combs, nursing brood, and storing honey. Finally, when a mature stage is achieved, with most of the brood sealed, queen cells are constructed at the lower rim of the comb, generally 12–16 in number. Now the bees stop enlarging the comb and rearing new brood. Swarming may announce the approaching end of an individual colony. By the short life cycle of a colony one is reminded of the fast-growing and early-flowering and dying plants which are able to finish their whole cycle within a few months.

The first virgins seem to hatch before or shortly after the first swarm leaves, probably together with the mother queen (Tirgari 1971; Whitcombe 1984a). Immediate swarming preparations are marked by a general restlessness and repeated runs of workers and the queen on the surface of the bee curtain. Now every few days another swarm issues as young bees and queens hatch successively. While the first swarm takes with it about half of the colony, the following swarms become smaller and smaller until a little group of bees is left on the big comb, which leaves with the last queen or disperses. This means that the swarming period ends with the complete dissolution of the colony. Lindauer (1956, 1957) observed five swarms from an individual colony in Sri Lanka within 12 days; finally only 400–600 bees remained with the mother comb, until at last they dispersed. Taking into consideration the seasonal migrations, it is completely justified to call *A. florea* the "migratory bee", or, in Oman, the "Bedhouin bee" (Whitcombe 1984b).

Complete disintegration of the colony after one swarming period, however, is not the inevitable end of the annual life cycle, as shown by the recorded history of a strong colony in Oman (Whitcombe 1984a): 5 Jan – start of brood rearing after winter pause; 11 Feb – first drone brood; 10 March – prime swarm with old queen; 12 March – first young queen emerged; 16 March – second swarm; 25 March – eggs from young queen in old comb; 5 May – colony gained in strength, new drone brood (old drones still present); 16 May – open queen cells; 25 May – colony absconded after disturbance. Here, a young queen started new brood on the same comb, terminating a swarming cycle, and apparently a second cycle was about to start 2 months after the prime swarm.

The distance to the new nesting site of a swarm may bee only a few up to several 100 m (Tirgari 1971). Sometimes a swarm changes its site several times until comb building is started. Even in the dry climate of the Persian Gulf a second swarming period occurs in autumn.

In spite of this very short life history of a *florea* colony, the life span of the queen may be much longer. However, nothing is known about it. In the flight room, one queen was observed to live for 1 year.

Little is known, also, about the mating behavior of *A. florea*. Drone flight time in Sri Lanka and Thailand was reported to be rather short, between 12.00 and 14.30 h, which is much earlier than the two sympatric species *dorsata* and *cerana* (Fig. 1.3; Koeniger and Wyanagunasekara 1976; Akranatakul 1977). *Florea* drones are attracted to 9 ODA and to dead *mellifera* queens (Sannasi et al 1971), while extracts of the mandibular glands of *florea* queens attract *mellifera* drones.

If the laying queen is removed from the colony, emergency cells are constructed by modifying worker cells (Kshirsagar et al. 1980; Free and Williams 1979). The emergency cells of *A. florea*, mainly occuring in the center of the comb, are slightly shrunk beneath the comb surface and therefore hard to locate (Fig. 7.22; Sakagami and Yoshikawa 1973). Evidently, mated replacement queens

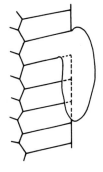

Fig. 7.22 Emergency queen cell on a *florea* comb (Sakagami and Yoshikawa 1973)

can be obtained in the queenless part of an artificially divided colony. In queenless colonies ovaries develop in worker bees followed by egg laying (several eggs in a cell; AkratanaKul 1977).

7.6 Genetics

The genetic difference from the other *Apis* species was analyzed for the amino acid sequence of melittin (p.24; Kreil 1973, 1975). Five residues are different in *A.florea* (out of 26) compared with *mellifera- cerana* and also five (but partly different) if compared with *A.dorsata*, indicating a fairly wide genetic distance.

7.7 Geographic Variability

Buttel-Reepen (1906) recognizes, in his synopsis of the genus, three varieties of *A.florea*: *andreniformis, florea,* and *rufiventris,* mainly discriminated by color (though Drory 1888 mentions the variability of color within one and the same colony). Maa (1953) reduced the number of varieties (=species in his special system) to two, a smaller *Micrapis andreniformis* (F.Smith) from submountain regions of Sri Lanka, Malayan Peninsula, Thailand, Sumatra, Java, and Borneo, and a larger *Micrapis florea* Fabr. from the lowlands of Oman, India, Sri Lanka, Malayan Peninsula, Sumatra, Java, Borneo, and Palawan.

In Yunnan, *andreniformis* occurs at lower altitudes (below 1000m), and in warmer climate. The pigmentation of workers is darker and the "thumb" of barsitarsus 3 in drones distinctly shorter than in *florea* (Fig. 7.23; Wu Yanru and Kuang Bangyu 1986); esterase isozymes in both types differ (Li Shaowen et al.1987) Therefore, *andreniformis* could well be regarded as a distinct species, although information on morphology, biology, distribution, and range of variation is still scanty.

At the Biometric Laboratory in Oberursel only 18 samples (360 bees of five regions (Sri Lanka, Thailand, Pakistan, Iran, Oman) were measured as a first attempt at morphometrics in *A.florea*. This is not sufficient for a comprehensive

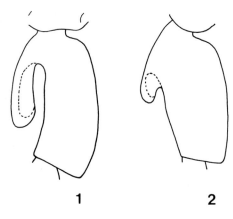

1   **2**

Fig. 7.23 Basitarsus 3 of drones of *A.florea* (left) and *A.andreniformis* (right) (Wu Yanru and Kuang Bangyu 1986)

Table 7.3 Some characteristics of *florea* bees of different origin. Measurements in mm

Origin	n (bees)	Proboscis	Fore Wing		3rd leg	Tergite 3 + 4	no hooks	Cub. index
			length	width				
South Iran	60	3.369	6.706	2.313	5.431	2.847	11.37	2.89
Oman	60	3.351	6.516	2.248	5.260	2.734	13.20	3.08
Pakistan (Peshavar)	40	–	6.598	2.316	5.202	2.738	–	2.85
Sri Lanka	80	3.156	6.168	2.125	5.118	2.630	11.60	3.50
South India	20	3.112	6.252	2.140	4.901	2.553	10.50	3.32
Thailand	60	–	6.433	2.201	5.269	2.748	–	2.94

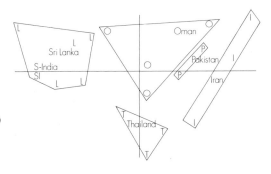

Fig. 7.24 Factor analysis (20 characters) of 18 samples of *A. florea*. Three groups are discriminated: Sri Lanka + S India – Thailand – Oman + Pakistan + Iran

analysis. However, it can be demonstrated that a clear geographic variability exists in *A. florea* even when using the same characters as in *A. mellifera*, with the same trend of variation as in the other species (Table 7.3): larger bees in the north and smaller ones in the south. Three groups can be recognized in PCA: Sri Lanka and S India, Thailand and Oman, Pakistan, Iran (Fig. 7.24).

7.8 Pathology

Since the economic significance of *A. florea* is not yet acknowledged, enemies and diseases of this bee have scarcely been investigated, and if so, only out of entomological curiosity. Yet, in *A. florea* also a species-specific parasitic mite was detected, namely the dermanysside *Euvarroa sinhai* (Delfinado and Baker 1974). No obvious damage to the colony by this mite was reported.

Quite accidentally, a new type of Nosema spores, possibly representing a new species, was detected while a *A. florea* colony originating from Sri Lanka was kept in flight rooms at the Institute in Oberursel. The spores are distinctly smaller and shorter than those of *Nosema apis* Zander ($4.66 \pm 0.05 \times 2.49 \pm 0.026\mu$ compared to $5.60 \pm 0.05 \times 3.08 \pm 0.02\mu$). The volume of the spore is only about half of N. *apis* spores. They propagate readily also in *mellifera* bees, retaining their small size even after three consecutive cycles in *mellifera* colonies (D. Mautz in Böttcher et al. 1975). This marked difference in size is evidently genetically determined, indicating a separate species of this protozoan. The experiments were abruptly discontinued to avoid the danger of transmitting the parasite to *A. mellifera*.

Florea bees and their nest were observed to be attacked by hornets, ants, bees of other species, different mammals (*Tupaia*), and bee- eating birds (Seeley et al. 1982). *Florea* combs may be destroyed by *Galleria* and other wax-feeding moths (Horne ans Smith 1870).

7.9 A. florea and Man

Several examples were presented in this chapter showing that man has not advanced beyond the level of a honey hunter and predator in his relations to *A. florea*. Nesting as it does in or close to settlements and practising mainly "passive defense behavior", it is an easy prey. The honey, though of low quantity, has a high reputation and, therefore, high prices. In the region of Bangkok, "bee hunter" is a well known profession and hundreds of *florea* nests may be "harvested" per week by one man and sold as a whole at the market (Akratanakul 1976, Fig. 7.18). It is proof of excellent adaptation to these adverse conditions and of high reproductive capacity, that *A. florea* is nevertheless able to survive. The population of *A. florea* in the three small towns of Dezful, Shushtar and Masjed Soleiman in Khusistan (Iran) was estimated to be 4000 colonies each in 1986 (Tirgari, pers. commun.).

There is, however, a unique example of "*florea* beekeeping" in the very restricted area of North Oman. *Florea* colonies are collected by experienced Arabs from the wild and, after removing the honey cap for consumption, the brood comb is sandwiched between two parts of a split stalk of a date palm leaf and located in wall niches or close to buildings (Dutton and Free 1979; Whitcombe 1984a). Honey is harvested without destroying the brood and sometimes colonies are transported during the winter to especially sunny places. Efforts are made to develop extended *florea* beekeeping, since this bee is better adapted to the special climatic conditions than *A. mellifera*. The adaptability to a new environment is shown by a recent report on occurrence of *A. florea* in the town of Khartoum, Sudan (autumn of 1985). *Florea* colonies thrive in feral conditions while the few *mellifera* colonies in the town, kept at the Bee Unit of the University, constantly have to be provided with food and water in order to survive (Bradbear 1986).

The most promising prospect for the function of *A. florea* in relation to man is pollination of crops in hot, dry lowlands. *A. florea* is better adapted to this environment than other honeybee species; it is an industrious field bee, easy to handle, and when it migrates, the flight range of the swarm is generally not very large.

CHAPTER 8

Apis dorsata Fabricius 1793:328

8.1 Introduction

The Giant honeybee is certainly the most spectacular of all four honeybee species: an individual bee of the length of a hornet, living in the open in huge colonies, frequently in exposed position (Fig. 8.1), the motionless bees with spread wings on the surface of the cluster arranged in strict regularity (Fig. 8.2) yet ready at any time to lance fierce mass attacks against a supposed enemy within seconds. While the other species live more or less hidden in cavities or dense bushes, *A. dorsata* evidently relies on its strength based on a numerous society of large individuals with high defense potential (Seeley et al. 1982). Yet, this "most ferocious stinging insect on earth" (Morse and Laigo 1969) can be conditioned to live close to humans, nesting on walls of buildings in large towns or on trees in gardens (Lindauer 1956; Morse and Benton 1967; Reddy 1980b).

Fig. 8.1 "Bee tree" in Sri Lanka with ten clearly visible *dorsata* nests. (Photo N. Koeniger)

Fig. 8.2 Nest of *A. dorsata* with "protective curtain". Wings of workers characteristically spread out. (Photo N. Koeniger)

Experiments with *A. dorsata* call for precaution and experience; a limited amount of data about the biology of the species has only been obtained during the last few decades.

8.2 Distribution

The area of *A. dorsata* covers almost exactly the Indo-Malayan region, as does *A. florea*. The distribution of both species, however, is not identical (Fig. 8.3): to the west, *dorsata* occurs not farther than the Indus river, avoiding the xerotherm

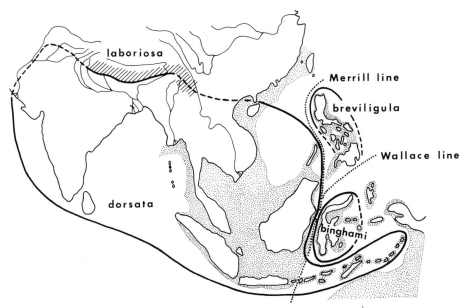

Fig. 8.3 Distribution of *A. dorsata* in S Asia with subspecies *binghami* and *breviligula* and the not yet definitely classified *laboriosa*. (Sakagami et al. 1980)

coasts of the Persian Gulf. To the east, however, all the Philippine islands are included in the *dorsata*-area which even crosses the Wallace line as far as the Kei Islands east of Timor (Fig. 8.3 Maa 1953; Sakagami et al. 1980). The Giant bee is reported to occur in altitudes up to 1000 m, to 1500–1700 m or, during migration, even up to 2000 m in different regions (Husain 1938; Muttoo 1956; Reddy 1980b; Gautam 1984). It seems to be common in areas where food and nesting sites are available. The number of colonies existing in Pakistan was estimated to be 50,000 (Ahmad 1984).

An interesting phenomenon is an extremely large variety, *"Apis laboriosa"*, frequently classified as an individual species, which is able to nest in the rocks of the Himalayan valleys in altitudes of 2000 to well above 3000 m, where nocturnal temperatures frequently fall below the freezing point (p. 118; Sakagami et al. 1980; Roubik et al. 1985).

8.3 Morphology

8.3.1 General

The most striking character on first sight is the rusty brown heavy pubescence and the dark "smoky" tinge of the wings (Fig. 8.4).

The general attribute "giant" usually applied to this bee can be accepted only for the dimensions of length, but not of width. If compared with *A. m. mellifera*, the bee next in size to *A. dorsata*, the wing length of *A. dorsata* (from Peshawar,

Fig. 8.4 *A. dorsata* worker

Table 8.1 Discriminant characters of the three subspecies of *A. dorsata* (Maa 1953)

Character	Tongue[a] length	Fore wing length	POL :	OOL :	LOL[b]	Jugo-Vannal-Index
Dorsata	5.95	12.5–13.0	12:	13:	6	148
binghami	6.27	14.0–14.5	11:	8:	7	133
breviligula	4.65	13.0–14.0	13:	9:	7	141

[a] It is not quite clear how the "tongue" was measured.
[b] See Fig. 4.6.

Pakistan) is $+31.5\%$, length of abdomen $(T3+4)+21.3\%$ (Table 8.1), but width of thorax is about the same in both species and the width of abdominal sterna is even smaller in *A. dorsata*, as shown in Fig. 8.5. The index of sternite 6, therefore, is raised to 102.1 compared to 78.6 in *A. m. mellifera*. The "enormous" tongue length of *A. dorsata*, stimulating the phantasy of beekeepers in the early days of exploring exotic bees (see p. 119), was reduced by morphometric investigation to a mere 6.68 mm in India (Rahman and Singh 1950), resp. 6.45 mm in Pakistan (pers. data).

The differences in size among the castes are very small, even less than in *A. cerana*; this is in strong contrast to *A. florea* (compare Fig. 8.6 with Fig. 7.2). Wing length in drones, compared to workers, is greater by 8%, but head width is even smaller (Table 1.3; Seeley et al. 1982). This explains the frequently mentioned fact that no specific drone size cells, but a uniform cell type is constructed. The sealed drone cell, however, shows a clear difference: it has an elevated, flat capping (Tha-

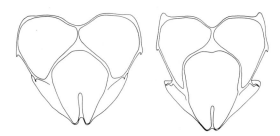

Fig. 8.5 Sternite 6 of *A. mellifera* (*left*) and *A. dorsata* (*right*)

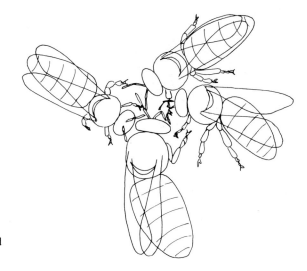

Fig. 8.6 *Dorsata* queen (*bottom*) with three workers. (Koeniger and Koeniger 1980)

kar and Tonapi 1961). Since no queen cells were known in early times, *A. dorsata* was believed to be a "uniform-cell-type" bee like Meliponinae, and hence more primitive (Buttel-Reepen 1903; Alpatov 1938) with the additional argument that *dorsata* workers show more ovarioles than the other species. This question is discussed in Chap. 1. The queen, too, is not much larger than the worker, except for the broad thorax (Fig. 8.6). This was taken as a possible adaptation to the migratory habit of this species, which makes it advantageous for the colony (including the queen) to take to the wing at any time (Koeniger and Koeniger 1980).

8.3.2 Genital Organs

Queen. The only investigation on female genitalia concerns the number of ovarioles which was found to be 130 per ovary (compared to 180–200 in *A. mellifera*; Velthuis et al. 1971). No data on the diameter of the spermatheca or counts of the spermatozoa stored are available.

Drone. The endophallus in situ was described by Bährmann (1961a), the everted organ by Simpson (1971). The bizarre structure of this organ, a unique speciality of Apini, is at its peak in this species. While honeybees of the female caste seem

Fig. 8.7 *Dorsata* drone with everted endophallus

rational and efficient in their morphology and behavior, all the free phantasy of nature apparently is concentrated on males. The main parts of the endophallus can be identified in all four species, but the total organ is much larger in *dorsata*, including the two bursal cornua and the bulb with its additional membranous protuberances (Chap. 3, Fig. 3.5). Even more conspicuous are three dorsal cornua, one of them, very long and thin, being sharply bent downwards (Fig. 8.7).

A further characteristic male *dorsata* organ which serves during copulation is the adhesive organ at the inner surface of basitarsus 3. It is a 0.5 mm thick brush (Fig. 8.8) consisting of robust, palm-like hairs (Fig. 8.9). The same is also found at the inner surface of the first three tarsal joints of the same leg (Fig. 8.8). This brush, which certainly increases adhesion to the queen's body during copula, is only faintly developed in drones of the multi-comb species *cerana* and *mellifera*.

8.4 Biology and Behavior

8.4.1 The Individual Bee

Development time is shorter in *A. dorsata* than in *A. mellifera* despite the somewhat lower nest temperature.: worker 16–20 d, queen 13–13.5 d, drone 20–23.5 d (Qayyum and Nabi 1968).

Fig. 8.8 Hair pads at the inner surface of basitarsus and tarsi of drone hind leg ("adhesive organ", Ruttner 1975)

1mm

No observations are available about the life span of the individual bee, but long-living bees have to be postulated from the life history of the colony: In most regions the colonies migrate over long distances twice or four times a year. One such distance, ranging to about 200 km, with several stops, takes 4–6 weeks (see Sect. 8.4.4). Since the colony build-up has to be included in these activities, only long-living bees can guarantee the colony's annual cycle. Moreover, in certain regions there are inactive periods of several weeks without brood rearing and field activity. (Buttel-Reepen 1903). It would be of interest to explore whether season-dependent differences in the physiological make-up of the individual bee exist.

8.4.2 Nest and Nesting Site

The *dorsata* nest might be regarded as just another open-air, single-comb nest, but it shows significant differences from the *florea* nest. It is always fixed underneath a broad support such as a thick limb of a tree or the eaves of a building. The comb is started with a mid-rib. The cells are attached to the mid-rib secondarily (Fig. 8.10), their opening slightly directed upward. The principle of construction is identical to *A. cerana* and *A. mellifera*, but fundamentally different from *A. florea* (p. 87).

Fig. 8.9 Coarse hair of adhesive organ shaped like a palm (Ruttner 1975b)

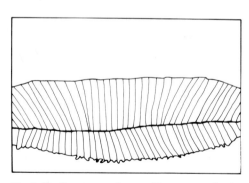

Fig. 8.10 *Dorsata* comb, cut off at support (horizontal section; drawn from a photo by J. Woyke)

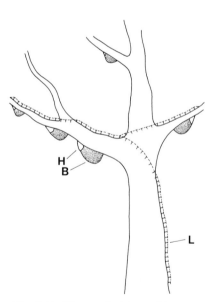

Fig. 8.11 "Bee tree" with primitive ladder (*L*) fixed to the trunk (Hadorn 1948). *B* brood section, *H* honey section of the comb

The shape of the comb is more or less semicircular or cuneiform (Fig. 8.11). The width of the comb (which corresponds to the line of attachment) is generally greater than the length. The surface of the comb varies according to the colony size from less than 0.1 m to well above 1 m². The comb size of 12 measured colonies in the Philippines varied from 110 to 8.120 cm² ($\bar{x} = 3.178 \pm 2.552$ cm²; Morse and Laigo 1969) with an average number of 23,300 cells. As in the other species the comb is differentiated from a thick honey comb in the highest section (Fig. 8.10) with a diameter of up to 15-20 cm, to a brood comb of 3.5 cm thickness (Roepke 1930; Hadorn 1948; Muttoo 1956; Morse and Laigo 1969; Seeley et al. 1982). Pollen is stored in cells between the two zones. The same cell can be used for storing or brood rearing, by simply extending or reducing the walls (Koeniger and Koeniger 1980). The uniform diameter of the cells is 5.35-5.64 mm, the depth of brood cells is 16 mm (Muttoo 1956; Thakar and Tonapi 1961).

In principle, this is exactly the pattern of a *mellifera* comb. By slowly widening the space between the combs in a multi-comb hive during a honey flow the thickness of the upper (storage) part of the comb with the honey can be considerably increased, but the depth of the brood cells remains unchanged. The *dorsata* comb is in every detail the same as that of a multi-comb colony, but without limitations of the depth of honey cells. The observation of a multi-comb *dorsata* nest by Morse and Laigo (1969) is of special interest (Fig. 8.12). Although the distance between the combs has not been sufficient to use the inner cells for brood rearing and this observation may be a very rare event, it is demonstrated that this nest exists within the species-specific range of variation and that transitions are found between the two nesting types of honeybees. According to a remark by F. Benton, multiple-comb colonies have been observed in nests under overhanging rocks (see Buttel-Reepen 1906).

The reports on the amount of honey stored in *dorsata* combs are highly contradictory. In general accounts the habitual formula seems to be "50-100 pounds" (= 22-45 kg). Whenever an exact survey was made, much lower figures were found: Kallapur (1950) investigated 170 colonies with an average amount of stored honey of 9 pounds (4 kg; range 2.8-16.9 lbs). Similar figures were found in the Philippines by Morse and Laigo (1969), which generally attested to a poor collecting tendency in *A. dorsata*. "Big colonies usually yield 10-15 kg honey if it is harvested at the right time" (Hadorn 1948, from Sumatra).

The stability of the big exposed combs seems to be a delicate problem. However, several factors have to be considered:

1. A large comb, 100 cm long and 20 cm thick, is attached to the limb by a surface of 2000 cm².
2. The honey is stored in the uppermost corner of the comb, close to the attachment.

Fig. 8.12 *A. dorsata* multi-comb nest in the Philippines: fixation area on tree limb. (Morse and Laigo 1969)

3. The weight of the bees (6500 per kg, that is 5-10 kg per colony; Muttoo 1956) is
   mainly held by the support, since the "protective curtain"(p. 104, 112) of bees is
   detached from the comb by 1-2 cm (Morse and Laigo 1969).
Nevertheless, indication of nests destroyed by typhoons were found.

The wax of *dorsata* combs has a lower melting point than *mellifera* beeswax and is
of special quality ("Ghedda wax"). The portion of wax esters in Ghedda wax is
85-90% compared to 71% in the wax of *A. mellifera* (Büll 1961). There are also differences in carbohydrate components (Hepburn 1986). Of the beeswax produced
in India 80% is of this kind (Phadke 1961).

   If a *dorsata* swarm decides to start a nest in a certain place, comb construction
is rather rapid. During one night a swarm constructed "a good portion of comb"
(Lindauer 1956; Koeniger and Koeniger 1980).

8.4.3 Organization of the Nest

Apart from the clear division of the comb into a honey part (with deep cells) and a
brood part, there is a functional organization of the bees living on the comb.

Nest Service. This is a thin layer of bees sitting right on the comb which are occupied with colony duties well known from *A. mellifera*: honey processing, comb construction, and brood care.

Protective Curtain. In a colony 80-90% of the bees are incorporated into a thick
"pelt" covering the comb (Morse and Laigo 1969). It is virtually a big bag consisting of three to six layers of uniformly oriented bees, fixed laterally to the support
of the comb, which is inserted into it. A bee space of 1-2 cm remains between
comb and curtain, serving as moving area for the "nest service", for ventilation
and temperature regulation. The individual bees of the curtain hang upwards,
wings spread, with an inclination of 5° towards the comb. Therefore, their position is exactly that of roofing tiles, with covered head and exposed abdomen
(Fig. 8.2). These bees are completely motionless unless disturbed. Irridescent wing
reflection probably contributes to temperature regulation. In the low part of the
curtain the position of the individuals is irregular, the heads are directed outward,
there is permanent movement of starting and landing field bees: the active area or
"mouth" of the colony (Fig. 8.13; Morse and Laigo 1969). In this area the communication dances are performed and attacks started. The active area is sharply
marked off from the rest of the curtain and its size varies greatly with the amount
of activity by the colony.

   Preferred *nesting sites* of *A. dorsata* are the limbs of big trees (from 5 cm to
100 cm; Seeley et al. 1982), buildings, monuments and cliffs in the mountains
("rock bee"). Nesting sites as high as 10 m and higher are generally found as part
of the defense strategy of these bees (Seeley et al. 1982), but this is by no means the
rule: nesting sites close to the ground, even if tall trees are close by, are not rare
(Morse and Laigo 1969). Differences may be due to different behavior of local
populations. Temporary sites during migration are usually close to the ground.

Fig. 8.13 *Dorsata* colony with clearly marked "mouth" in the lower part. (Photo Morse and Laigo 1969)

A phenomenon not completely understood is the aggregation of many (30–150) colonies on the same tree or building (Roepke 1930; Hadorn 1948; Lindauer 1956; Morse and Benton 1967; Koeniger and Koeniger 1980; Seeley et al. 1982). The last-named authors explain this by the availability of tall trees, because they found a more even distribution in the primary forests of Thailand than on cultivated land. Hadorn (1948), however, found heavy aggregations in Sumatra in spite of the presence of uncolonized trees. In the Philippines only solitary colonies were found in a region rather densely populated (about 1 colony/km^2 Morse and

113

Laigo 1969). One factor for aggregations certainly is the tendency to form nuclei (Lindauer 1956). Most likely it is the special character of the site which makes it attractive for bees. On the other hand, overcrowding must be dis-advantageous for provisioning. The possible common defense of the colonies of one tree was excluded by Seeley et al. (1982), who demonstrated that alerting one colony did not affect a neighboring colony at all.

A deserted comb is usually destroyed to a great extent during the absence of the bees. There is disagreement whether the remnants of old combs are used for constructing a new nest. Mahindre et al. (1977) and Reddy (1980b) affirm this, but Hadorn (1948) states that it would never occur.

8.4.4 Defense Behavior

Due to the size of the individual bee and the size of the population, *A. dorsata* is rarely - if ever - attacked by ants (the most dangerous enemies of smaller species). The very effective behavior fending off small intruders is the "shimmering" movement described in detail on p. 134 (*A. cerana*). Most small animals are repelled by this, but if one lands on the colony nevertheless, it is immediately mobbed by a number of bees and eliminated by a cluster of bees falling with the intruder to the ground. Of all insects which catch bees in flight, large wasps have the greatest impact on *dorsata* colonies. While an approaching *cerana* bee elicits only one to three waves of shimmering, there are about 50 when large *Vespa* species are approaching (Roepke 1930). Persistent attacks of wasps or birds spezialized in catching field bees (*Merops*) cause complete stop of flight activity (Koeniger and Koeniger 1980). The other specific defense strategy of *A. dorsata* is directed against large vertebrates like bears, birds, and man (*Pernis apivorus, Merops* and *Muscicapa*; van der Meer-Mohr 1932). A mass attack by a defense force of up to several thousand bees recruited from the thick "protective curtain" is the main tactic used. Therefore *dorsata* colonies possess a much larger force of potential defenders than the other species. A colony alarmed by a "disturbed" bee performing quick runs over the curtain forms clusters at the lower edge of the colony. The bees of the clusters take to the wing by dropping down and then start attack flights in groups close to the ground. One single bee initiates the flight of up to 5000 bees. Alarmed bees ready to attack were observed in large number at 1.6 km distance from the nest and a few even at 3.2 km. It took 2 days until the disturbed colony came to rest, reacting all the time nervously (Koeniger 1975). The alerting effect of *dorsata* sting extract persists two to five times as long (6-15 min) than extract of *mellifera* sting. This reveals another defense mechanism: long-lasting marking of an enemy by alerting pheromones. *A. dorsata* has the highest amount of isopentyl acetate of all *Apis* species, but no 2-heptanone (Morse et al. 1967). Instead, Koeniger et al. (1979) detected another sting pheromone, present also in *A. florea*: 2-decen-1-yl-acetate. Barbs on the dorsata sting are longer than in any other species (Weiss 1978).

Frequent alerting reactions which can be released to a lesser degree by incidents like a passing butterfly or a falling leaf are energy-consuming. Therefore, a process of habituation to repeated "safe" stimuli is of great advantage. This is what evidently happens to the *dorsata* populations in the big cities of India, where this bee is better tolerated than in other countries. Reddy (1980b) located

340 colonies in the town of New Delhi without finding a single report of ferociousness. Koeniger (1975) describes this process of habituation during experiments with *A. dorsata* in Sri Lanka. The first approach to the colony, located within a dense forest, has to be attempted during the night and very cautiously. A protective cabin is constructed close to the colony, and offers retreat. Within a few days the colony grew accustomed to the presence of the cabin and of moving objects, and later the space around the nest can be cleared without major attacks.

8.4.5 Regulation of Microclimate

Visvanathan (1950) and Morse and Laigo (1969) measured the temperature of the *dorsata* brood comb. They found a nest temperature fluctuating between 30 and 32° C, while the ambient temperature varied from 20 to 28° C. The protective curtain is certainly essential for maintaining a fairly constant brood temperature. It can be inferred from observations made with *A. florea* that the distance between comb and curtain is variable in order to regulate the microclimate of the nest. The reflection of radiation by the spread of wings on the surface of the cluster may have a thermoregulatory effect. No experiments are known about the efficiency of regulation at temperatures lower than 20° C, but the survival of a spezialized *dorsata* type *(A. d. laboriosa)* in a cool-temperate zone proves that the system of regulation can be developed to a high level of perfection. Shimmering waves periodically running over the curtain at ambient temperatures lower than 10°C may be understood to be heat–generating activity (Wolfgang Ruttner, pers. observ.).

8.4.6 Migration

Regular seasonal migrations are observed in almost all tropical *Apis* species or races, but rarely in an almost obligatory way as in *A. dorsata*. Differing from *A. florea*, there are mainly long-distance migrations, at the beginning of the dry season to the mountains, and of the rainy season to the plains and the coast. Frequency and timing of migration depends on the climatic rhythm of a region: in the uplands of Sri Lanka and in Bangalore (center of South India) the swarms arrive with the start of the dry period (Oct-Dec), and they leave (for the coastal plains) at the beginning of the Monsoon rains (May-July); (Koeniger and Koeniger 1980; Reddy 1980b). The migration distance in Sri Lanka is 150-200 km. In regions with two pronounced monsoon periods (east coast of Sumatra) two commutation cycles are observed, with a stay of only 2-3 months at the nesting site (Hadorn 1948). While preparing for migration the rearing of brood is stopped. No brood and honey were found in deserted combs. The departure is indicated by "migration dances" all over the cluster (Lindauer 1956; Koeniger and Koeniger 1980), pointing out the flight direction. The start of the swarm is shown by "buzz-runs" ("Schwirrläufe") of individual bees. Koeniger and Koeniger (1980) were able to follow some migrating swarms, at least partly. The distance to the new nesting site was accomplished in several stages. Stops of 1 to 3 days each were made on the way. The swarm settled generally close to the ground, no comb was constructed, but nectar gathering took place. The distance between stops was from 200 m to

5 km. The migration for a distance that could have been managed within 5-10 h of direct flight took 1 month.

The seasonal long-distance migration of *A. dorsata* poses as many unsolved questions as the migration of birds. At present, the mere description of the phenomenon has not progressed beyond some initial steps.

Environmental factors like shelter and availability of forage and water are given as causal agents. There are, however, strong indications for a genetic basis of this behavior: *dorsata* colonies kept in flight rooms in constant conditions (temperature, light-dark, food, humidity) stopped brood rearing every few months, then left the comb to settle in another corner of the cage, but finally resumed brood rearing activity (Koeniger 1977). One *dorsata* colony was kept for 26 months, with intermittent brood activity, in a flight room of the Institute at Oberursel.

8.4.7 Field Activities and Communication

All authors agree that *A. dorsata* nests only in locations where the sun or at least parts of the sky are visible. Bee trees always have sparse foliage (Fig. 8.1). Lindauer (1956) established that *A. dorsata* is able to transpose the horizontal direction to a food source to the vertical comb, in relation to gravity, but only during simultaneous sight of the sun or at least of the sky. Koeniger and Koeniger (1980) found that dances were performed only on the brightest spot of the comb which moves with the position of the sun in the course of the day. The decrease of frequency of wagtail runs with increasing distance is similar to *A. mellifera*, that is slower than in the other tropical species. This means that also greater distances can be communicated to the nest members. *Dorsata* honey is collected from fewer plant species than honeys from smaller species, indicating that this robust flyer is specialized in exploiting rich nectar sources even at distances farther than 5 km (Koeniger and Vorwohl 1979). However, so far nobody has been able to train *A. dorsata* to foraging distances greater than 400 m (Punchiheva et al. 1985).

A. dorsata was observed to forage and dance also during bright nights (Divan and Salvi 1965). The extrapolated position of the sun (Dyers 1985 b) is used for compass orientation, and not the moon. A proof of nocturnal activity is also the repeatedly observed attraction of *dorsata* bees to street lamps.

8.5 Reproduction

Queen Cells. It was believed for a long time that the queens are reared in the same cells as workers and drones (Buttel-Reepen 1903, Butler 1954). This supposedly low level of caste differentiation was used as the strongest argument for a primitive state in evolution. Later this turned out to be lack of knowledge: *A. dorsata* produces queen cells of the general *Apis* type on the lower edge of the comb, 5-11 in a row, partly immersed in the comb (Millen 1942b; Viswathan 1950; Thakar and Tonapi 1961; Morse and Laigo 1969; Reddy 1980b).

Drone brood is irregularly scattered among worker brood. Drones are reared in the same cells as workers, but the cappings are elevated. Adult drones are found aggregated in the upper part of the brood comb concealed underneath the curtain (Morse and Laigo 1969).

Reproductive Swarms. Lindauer (1956) described the process of reproductive swarming which differs from the sudden departure of a migratory swarm: during the "play flight" of young queens a line of bees flying back and forth in a certain direction is formed while the whole colony becomes agitated. Later fewer and fewer bees are seen on a return flight and the connection between mother colony and swarm is interrupted. No dances were observed. During the short swarming period small swarms depart every 3–4 days. One swarm was observed to settle at a distance of 500 m. Besides swarming, a process of "sprouting of a nucleus" is observed, unparalleled in the genus, but which has much in common with the act of swarming as described before and with the "progressive start" of new colonies in Meliponinae (A. Wille 1983): a gradual separation of a part of the colony. A small cluster is formed on the same limb of the tree about 1 m apart from the mother colony and a new comb is built. For several days many bees are seen comuting between the new and the mother colony until the daughter colony is completely self-supporting (Lindauer 1956).

Mating and Laying Workers. Not more is known about the mating biology than the very short flight period of drones - only for half an hour right before dusk (Fig. 1.3). A colony with young brood deprived of the queen is able to raise emergency cells. In a queenless colony laying workers appear soon (Millen 1942b; Morse and Laigo 1969). Ovaries of *dorsata* workers are large, with 15–60 ovarioles each (Alpatov 1938a; Velthuis et al. 1971). This certainly is a symptom of a lower level of social organization, corresponding to a high, queen-like level of yolk proteins in the worker hemolymph (Engels 1973).

8.6 Geographic Variability

It is evident that *A. dorsata* is surprisingly uniform in its main area. Although few morphometric data are available, the standard deviation of a sample of 71 bees collected from such distant locations as India, Nepal, Thailand, Cambodia, Laos, Vietnam, N Borneo, and Palawan was within the usual limits, as if originating from the same location (Sakagami et al. 1980). Even Maa (1953), always ready to create new taxa, did not split the species. This uniformity seems to exist also throughout the islands of SE Asia isolated only since the Holocene (see Chap. 3), where a number of differing island populations exist in *A. cerana* (Chap. 9). Provided the uniformity of continental and most *dorsata* island populations can be confirmed morphometrically, this can be taken as an additional argument for the immobility of *A. dorsata* in evolution (Chap. 3).

Two subspecies differing from the species type were described in border regions of the area, *A. d. binghami* and *breviligula* in the east (Table 8.1), beyond the Wallace-Meryll line (Fig. 8.3). Both the areas of *binghami* in Celebes and of *breviligula* in the Philippines were more or less isolated from the continent during the Pleistocene (Sakagami et al. 1980).

Apis dorsata breviligula (Maa 1953, p. 563) seems to be characterized by an extremely short tongue, and by medium length of fore wing. Other characteristics given by Maa are hard to quantify. The detailed description of the biology of this subspecies given by Morse and Laigo (1969) shows the "key" *dorsata* characteristics, but also some differences. Congregations of several nests on the same support, common, e.g., in India (Reddy 1980b), was never observed in the Philippines.

A. d. binghami Cockerell(1906) has the longest tongue and forewing. The tomenta are described as snow-white and narrow, the cover hair black, ocelli as comparatively large (Maa 1953). Area: Celebes, Sula Islands.

The fourth type belonging to the *dorsata* group is *A. d. laboriosa*. The taxonomic problems involved are discussed in detail in Chap. 1, together with morphological characteristics of "the world's largest honeybee" (Sakagami et al 1980). Only a limited number of specimens of this specialized mountain type were collected so far in valleys of Nepal, Butan, Sikkim, and Yunnan at altitudes between 1200 and 4100 m with a maximum frequency between 3000 and 3500 m (evergreen oak zone with cool-temperate climate; Sakagami et al. 1980; Roubik et al. 1985). In this zone the "only nontropical honeybee" experiences ambient temperature regimes between $10°$ and $-5°C$ or colder during all but a few months. Flight activity should be limited to a few months in summer if temperature requirements are the same as in *A. mellifera*, but active *laboriosa* field bees have been observed during Oct-Dec.

Roubik et al. (1985) reported nest aggregations on south-facing, steep cliffs of deep river valleys in altitudes of 2600-3200 m. There is no clear evidence whether the colonies migrate to lower altitudes during the cool season or not. A clear ecological isolation from *A. d. dorsatao* seems to exist. *A. d. laboriosa* evidently represents the most specialized of all open-air-nesting honeybees. The capacity of microclimate regulation at cool ambient temperatures of an open-air-nesting colony is a quite remarkable accomplishment. It compares well with the thermoregulation of an *A. mellifera* swarm in cool temperatures (Heinrich 1985). An observation communicated personally to the author by W. Ruttner may be interpreted as a step in the direction of a multi-comb colony: beneath an eaves-like rock in Nepal (alt. 3000 m) several *labriosa* colonies were nesting. Some of them were arranged in groups of three to five combs each closely and regularly spaced together, but the clusters remaining separated. This "multi-colony" arrangement must help thermoregulation (there was snow on the opposite shaded slope from recent snow fall). *A. (d.) laboriosa* seems to indicate a different line of evolution in Apini, pointing rather to Bombinae than to cavity-nesting *Apis* species: large body size, dense and long hair, dark color. Many questions about the physiology of this bee, mostly nesting in inaccessible sites, remain open.

8.7 Pathology

A. dorsata has its own species-specific parasites, as have the other species. Two species of mites were found, *Tropilaelaps clarae* and *T. koenigerum*, evidently occurring all over the *dorsata* area (Delfinado 1963; Koeniger and Delfinado-Baker 1983; Delfinado-Baker et al. 1985). *Tropilaelaps* is sometimes found also in colonies of the sympatric *A. cerana* (Delfinado-Baker 1982), but it commuted with disastrous effect to imported *A. mellifera* colonies (Koeniger 1982; Woyke 1984). *A. dorsata* is only to a minor degree affected by *Nosema apis* (Koeniger 1977).

Wax moths, a common pest for all *Apis* species, also have an impact on *A. dorsata*. Of 272 colonies examined in India, 90 were heavily infested. Wax moths may become fatal to the colony during periods of dearth and weakness and/or queenlessness (Sihag 1982 b).

8.8 Apis dorsata and Man

Early attempts were made to hive *A. dorsata* and to transport colonies to countries of the temperate zone (see Buttel-Reepen 1906; Morse 1970 a): Dathe succeeded in bringing a living colony to Germany in 1883, Benton shipped four colonies to Syria (1899) in the attempt to introduce the Giant bee to the USA and in the same year Jones brought a colony to Australia. More than 70 years later Thakar (1973) published experiments with a "hive" for *A. dorsata*.

The high credit given to *A. dorsata* as a honey collector is derived from the fact that 60-70% of the honey and virtually all of the wax harvested in India is taken from *dorsata* colonies (Thakar and Tonapi 1961; Singh 1980). The honey is collected during the night by trained people who climb the tree (Fig. 8.11). The bees are driven off the comb by smoke and fire. Generally only the honey section is taken while the brood remains untouched (Hadorn 1948; Schmidt et al. 1985). This extensive predation, however, seems to be of less impact on the species than the destruction of the primary forests with its tall trees.

A. dorsata is listed as an important pollinator in several reports (Kapil, Atwal (1970); Maun and Gurdip Singh 1983).

CHAPTER 9

Apis cerana Fabricius 1793:327

Apis indica Fabricius was used as synonym for this species, at least in the western part of its area. The correct taxonomic name, however, is *Apis cerana*, which was assigned by Fabricius in 1793 for a bee in China. Later, in 1798, the same author named honeybees of India *A. indica*. Since there is general agreement that both geographic types belong to one and the same species, the term *Apis cerana* has priority (Lindauer and Kerr 1960).

9.1 Introduction

Apis cerana, the "Eastern Honeybee" (Butler 1954), is the exact equivalent, in the Eastern part of the Old World, of its occidental sister species *A. mellifera*. It has an equally wide area of distribution with a similar capacity for a broad spectrum of adaptations.

In the north, *A. cerana* occurs up to 46° northern latitude in Ussuria compared to 58° at the northernmost limit of the original distribution area of *A. mellifera* in South Scandinavia and 60° in the USSR (Leningrad). However, in Ussuria *A. cerana* meets with much lower winter temperatures, comparable to those in the southern Ural (Bashkiria) at 50° northern latitude where *A. mellifera* exists in feral colonies. *A. cerana* is indigenuous in the tropical rain forest in Malaysia, as is *A. mellifera* on the banks of the Congo. *Cerana* was imported to Hawaii from East Asia (B. Schricker, pers. commun.), as was *mellifera* to S America. Nests are found well above 2000 m and even at 3000 m in the southern valleys of the Hindukush and the Himalaya (Mattu and Verma 1980), corresponding to the same altitudes for *A. mellifera* in the mountains of East Africa (Smith 1961). *A. cerana* shows the same ability of adaptation to a very dry, semi-desertlike environment in the mountains of Central Afghanistan or in the plains of India, as does *A. mellifera* on the Arab peninsula or in the oases of the Sahara.

9.2 Area of Distribution

Apis cerana occurs in all of Asia east of Iran and south of the great mountain ranges and the central deserts (Fig. 9.1). In the north, there was originally a gap of 5300 km where no bees were found between the eastern border of *A. mellifera* in the Ural and the northernmost occurrence of *A. cerana* in Ussuria. In Siberia *mellifera* bees were introduced only during the 19th century (Skorikov 1929b).

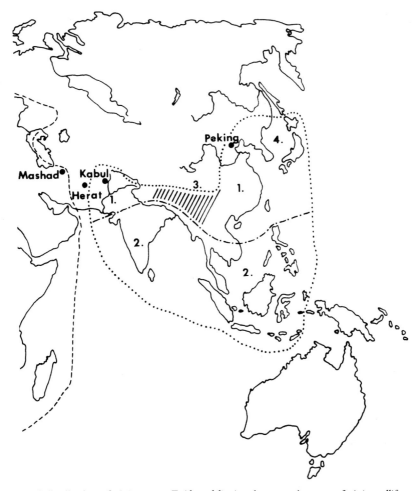

Fig. 9.1 Area of distribution of *Apis cerana* F. *(dotted line)* and gap to the area of *Apis mellifera* L. *(broken line)* (1) *A.c.cerana*; (2) *A.c.indica*; (3) *A.c.* Himalaya; (4) *A.c.japonica*

In the east *A. cerana* was found in feral conditions as far north as the district of Primorje at 46° (Lavrekhin 1958, Fig. 9.2). In Japan, *A. cerana* is evidently missing in the northern island of Hokkaido, while it is widely distributed over the other islands, including Honshu (Sakai and Matsuka 1982). Okada (1970) established the presence of cerana colonies in the Shimokira peninsula in the north of Honshu (Fig. 9.3). In Southeast Asia *A. cerana* is restricted to the Malayan region, west of the Wallace line (Philippines – Celebes – Timor), occurring in different island races (Maa 1953). On the Moluku Islands (Buru, Obi, Batjan) no honeybees were found except on Ambon. Here, however, bees supposedly had been imported about 150 years ago (Roepke 1930; Maa 1953).

It was of special importance to ascertain the exact border of the *cerana* area in the west. Several authors reported on transitory types between *A. cerana* and *A. mellifera* in Kashmir (Vats 1953; Deodikar and Thakar 1966; A. M. Shah 1980).

121

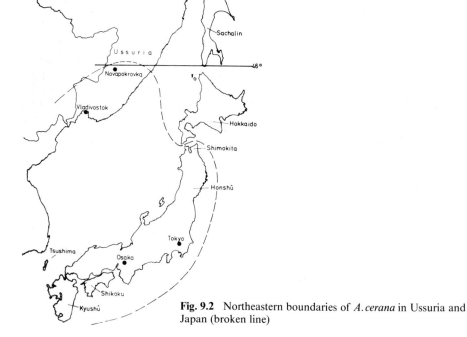

Fig. 9.2 Northeastern boundaries of *A. cerana* in Ussuria and Japan (broken line)

G. Nogge (1974) was able to locate the westernmost colonies of *A. cerana* in West Afghanistan. He discovered bee hives in the province of Ghor in the village Sar-i-Ghor Mushkan at an altitude of 2300 m. Beekeeping is a tradition there and feral colonies were said to occur as well. The bees collected in this region were determined to be pure *A. cerana* without any hint of transition to *A. mellifera* (see p. 153 Fig. 9.20). The location is 240 km southwest of Herat (Fig. 9.4).

In the Koh-e-Binalud mountains near Mashad, Iran, about 360 km northwest of Herat, pure *mellifera* colonies were discovered by the same author and by Ruttner et al. (1985b). The occurrence of *mellifera* or *cerana* bees in East or Southeast Iran (Khorasan, Baluchestan) has thus far not been verified (Pourasghar, pers. commun.)

Thus a belt of deserts separates *A. cerana* and *A. mellifera*. However, at the northern sector of the border between Iran and Afghanistan the two species occur only 600 km apart. It was demonstrated by the above investigations that *A. mellifera* and *A. cerana* are two strictly allopatric species (Ruttner and Maul 1983). No overlapping or direct contact between the two areas of distribution was observed (Ruttner et al. 1985b).

9.3 Morphology

Apart from a few conspicuous qualitative characteristcs which are mainly used to discriminate between the very similar species *A. cerana* and *A. mellifer*a, e.g., the radial vein of the hind wing, the tomentum on the 6th tergite or the absence of

122

Fig. 9.3 Feral colony in a hollow tree in Gamanosawa, N Honshu. (Photo I. Okada)

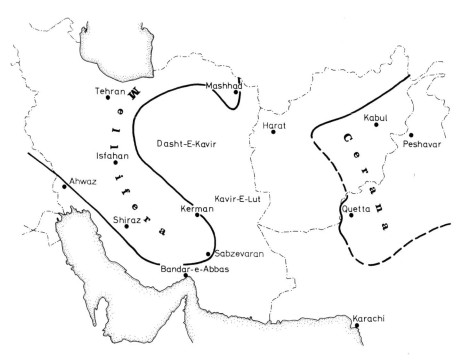

Fig. 9.4 Eastern boundary of *A. mellifera* in Iran and western boundary of *A. cerana* in Afghanistan and Pakistan. (Ruttner et al. 1985b)

chitinous plates on the endophallus (Fig. 9.9, Fig. 9.11), others are found which are also specific for *A. cerana*. A summary of a number of differences in both castes is given in Tables 1.1 and 9.1 illustrated by Figs. 9.5–9.13. Other morphological differences will be mentioned elsewhere in this chapter. Two quantitative differences, cubital index and number of hooks on the hind wing, are true species-specific variables, with almost no overlapping (Fig. 9.10). One quantitative difference, *size*, needs detailed discussion.

The "small body size" of *A. cerana* was alleged by many authors to be the major discriminant factor. However, this assumption favored misidentifications, e.g., the allocation of "*Apis cerana*" races to Senegal and Cameroun (Buttel-Reepen 1906; Enderlein 1906; Maa 1953).

The main reason for this misinterpretation is the overlapping of the variabilities in size of both species (Fig. 4.4, Fig. 9.12). Varieties occuring in the plains are smaller in both species than those of the mountains; the same is true for the varieties of the tropics and the temperate zones in the two species. However, comparisons were generally made between large European races of *Apis mellifera* known everywhere (mostly *ligustica)* and relatively small southern *cerana* types, instead of comparing races of the same geographic latitude. Small southern races of *A. mellifera*, like *yemenitica* or *adansonii*, are even smaller than medium-sized *cerana* races (Fig. 4.2, 4.4), and there is no significant difference between the smallest races of both species: the big races of *A. mellifera*, however, e.g., those of Europe, considerably exceed the biggest *cerana* races in size (Fig. 4.4). The overall mean values of body dimensions in the species *A. cerana*, therefore, are smaller than in *A. mellifera*.

Table 9.1 Selection of several species specific characters of *A. cerana* and *A. mellifera*. For the three major species-characters of *A. cerana* see Table 1.1

No.	Character	A. cerana	A. mellifera	Author	Fig.
1	Labrum – pigmentation	all yellow or brown	All dark or dark with yellow marks	Bährmann	9.5
2	Labial palpus – length of segm. 3 : Length segm. 4	1:1.3	1:1	Enderlein 1906	9.6
3	Index of wax mirror (mean value)	51.8	56.0	Alpatow 1948	9.7
4	Tibia of hind leg – drone	Grove (longitud.)	Round	Bährmann	9.8
5	Tomentum index (mean value)	0.85	1.95	Ruttner (unpubl.)	9.9
6	Cubital index (mean value)	4.40	2.30	Goetze 1940	9.10
7	Hooks on hind wing (mean value)	18.28	21.30	Goetze 1940	9.10
8	Endophallus Chitinous plates	Absent	Present	Simpson 1960	9.11
	Upper cornua	Three pairs	Rudimentary	Ruttner et	9.11
	Fimbirate lobe	Rosette-like	Feather-like	al. 1973	9.11

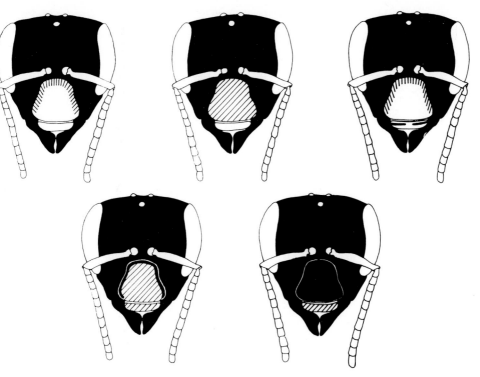

Fig. 9.5 Head of *A. cerana* – pigmentation of frontal part. Labrum: all yellow or brown, in some cases yellow with dark stripe. Clypeus: yellow with brown border, all brown (sometimes with darker border) or all dark

a   b

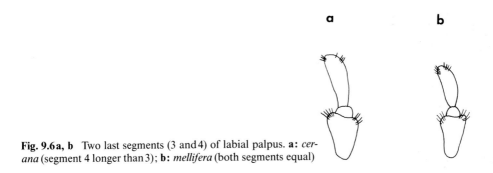

Fig. 9.6a, b Two last segments (3 and 4) of labial palpus. **a:** *cerana* (segment 4 longer than 3); **b:** *mellifera* (both segments equal)

Genitalia

The differences between *A. cerana* and A. *mellifera* are most striking in the male genitalia (Fig. 9.11), while they are easily overlooked in the female castes. This may help to explain the long-lasting incertitudes regarding both taxa in respect to each other, since male genitalia were frequently not examined. An interesting characteristic of the endophallus are the three pairs of dorsal cornua. They are developed in

125

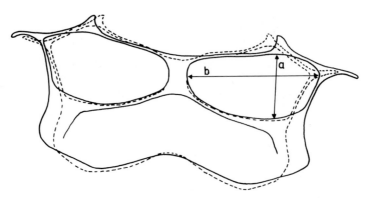

Fig. 9.7 Sternite 3. *Solid line: A.cerana. Broken line: A.mellifera* (Index of wax mirror = (a:b) · 100

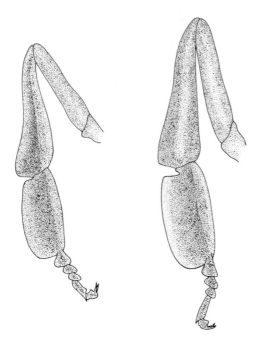

Fig. 9.8 Hind tibia of drone. *Left: cerana; right: mellifer*

Apis dorsata into long, thin appendices (Fig. 8.7) while in *A. mellifera* they exist only as hardly recognizable rudiments (Fig. 3.5). Thus *A. cerana* appears to be in the middle in the line of evolution of this structure.

9.4 Biology and Behavior

Biological data are much more difficult to collect than specimens of dead bees. Thus what has been said about "species characteristics" in morphology (e.g., size) applies especially to this section. Is a certain observation about *Apis cerana* from

Fig. 9.9 *Cerana* worker

Table 9.2 Numbers of bees of *A. cerana* and *A. mellifera* that moved or were still after exposure to low temperatures for 1 h. (Free and Spencer-Booth 1961)

Environmental temperature (° C)	*A. cerana* Moved	Still	*A. mellifera* Moved	Still
8	0	12	0	12
9	0	17	3	15
10	1	30	20	10
11	22	22	33	10
12	15	12	21	3
13	21	4	26	0
14	6	0	6	0

one region only of local significance or is it characteristic for the whole species? Or could it be even a characteristic transgressing the species, not yet recognized as the restricted starting point of European *mellifera* races?

A good example for the need of very cautious interpretations is a special type of swarming behavior. Formerly "absconding" (desertion of the whole colony) was described as a typical behavioral trait of *A. cerana* (Tokuda 1924, 1935;

127

Cubital-Index

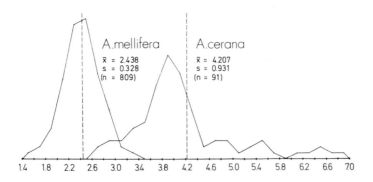

A.mellifera
$\bar{x} = 2.438$
s = 0.328
(n = 809)

A.cerana
$\bar{x} = 4.207$
s = 0.931
(n = 91)

1.4 1.8 2.2 2.6 3.0 3.4 3.8 4.2 4.6 5.0 5.4 5.8 6.2 6.6 7.0

Hooks

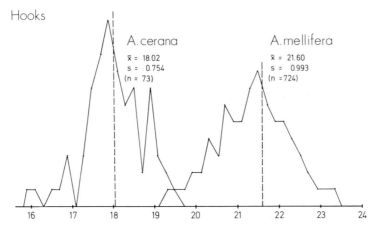

A.cerana
$\bar{x} = 18.02$
s = 0.754
(n = 73)

A.mellifera
$\bar{x} = 21.60$
s = 0.993
(n =724)

16   17   18   19   20   21   22   23   24

Fig. 9.10 Distribution of cubital index ($\bar{x}$ of samples), *left: mellifera; right: cerana*. Bottom: frequency distribution of number of hooks on hind wing, ($\bar{x}$ of samples). *right: A.mellifera; left: A.cerana*

Roepke 1930; Kellog 1941; Sakagami 1960a). Butler (1954) reported that the two other honeybee species of SE Asia, *Apis dorsata* and *Apis florea*, show the same phenomenon. Moreover, *Apis mellifera* races of tropical Africa also frequently abscond, and for the same reasons (irritation, lack of food, seasonal migration) as does *Apis cerana* (Smith 1961; Fletscher 1978a). Absconding is obviously an important behavioral adaptation to tropical conditions in all *Apis* species (Lindauer 1956). On the other hand, other characteristics like position during fanning or the pore in the capping of drone cells seem to be strictly species-specific.

It will have to be examined in every single case how far so-called "*cerana*-specific" phenomena can be generalized (or restricted) to the whole species as far as present knowledge is concerned.

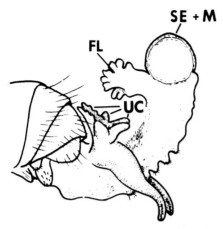

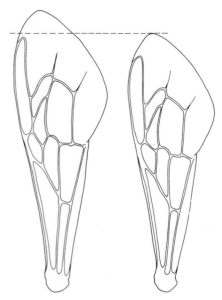

Fig. 9.11 Everted endophallus of *cerana* drone. *FL* fimbriate lobe; – *UC* upper cornua; *Se* mucus and semen

Fig. 9.12 Length of fore wing of *A. cerana* from Peshawar (34 n lat.) *left*, and of *A. mellifera* from North Oman (23° n lat.) *right*

9.4.1 Physiology of the Individual Bee

No fundamental differences can be expected in the individual physiology compared to *A. mellifera*, since the ecological conditions are very similar. Nevertheless, some divergences have been recorded.

9.4.1.1 Time of Development

For *cerana* worker bees in India the embryonic development took 74 h, the larval development (to capping) 196 h (= 8.2 days). The capped cell stage lasted 11 days, thus giving a total of 19,2 days for the development from the laid egg to the adult (Mishra and Dogra 1983; Rahman 1945). This is less than the time required for European races, but equals tropical *mellifera* races (p. 202, 211, 227).

9.4.1.2 Respiration Metabolism

Oxygen consumption in µl per mg live weight per h (Warburg respirometer at 32°) was higher initially in Japanese than in European worker larvae (5.74, 4.03), but was about the same at the end of larval growth. Oxygen consumption was generally higher in European than in Japanese drone larvae. In adult bees oxygen consumption increased with age. At 20 and 30 days the respective rates were 6.60 and 11.56 for Japanese and 4.74, and 6.51 for European bees (Hukusima and Inoue

1964). These differences in oxygen consumption of adults might be explained by the higher irritability of *A. cerana*.

9.4.1.3 Enzymes

In their hemolymph *A. mellifera* and *A. cerana* (from Japan) have the same four esterase isozymes, but the bands of *A. c. japonica* migrate much slower than those of *A. m. ligustica* (Tanabe and Tamaki 1986). For enzymes added to the nectar see Section 9.4.5.

9.4.1.4 Temperature Tolerance

Groups of 100 *cerana* workers from Sri Lanka, kept in cages, showed the same level of temperature at the center of the group and the same capacity to react to varying environmental temperatures as European (supposedly English) bees (Free and Spencer-Booth 1961). However, mortality was higher in *cerana* bees at temperatures up to 25° C. With increasing environmental temperatures water consumption rose and food intake decreased in both species. Oxygen consumption of young *cerana* bees from Punjab, India, was about the same (with temperatures ranging from 5°C to 45° C) as that of European *mellifera* bees. At 50° C, however, *cerana* bees survived for a much shorter time, whereas they equalled *A. mellifera*'s survival time at 5°C (Verma and Edwards 1971). Chill coma started at temperatures about 3°C higher in the bees of Sri Lanka than in *mellifera* of England (Table 9.1). Again it has to be noted that bees of the tropics had been compared to those of the temperate zone.

Some data on temperature regulation in full *cerana* colonies published by Kapil (1960) and L. R. Verma (1970) indicate that thermohomeostasis is also highly developed in *A. cerana*. At an ambient high of 41°C the temperature was 3 degrees higher in a *cerana* hive than in a *mellifera* hive, apparently due to a lower evaporation rate. In winter (ambient temperature 2° C), *A. cerana* clustered more compactly with cluster temperature of 37°C compared to 34°C cluster temperature for *A. mellifera*. Compared to exact data available for *A. mellifera*, however, no detailed investigations on temperature regulation exist for *A. cerana* colonies. Several corresponding reports on comparative observations state that *cerana* bees fly at 2–5°C lower temperatures (earlier in the morning or season) than *mellifera* bees (Ankinovich 1957; Lavrekhin 1958; Danilova 1960; L. R. Verma 1970; S. M. Shah 1980; O. P. Sharma et al 1980; and others).

9.4.1.5 Wing Beat Frequencies

A. cerana from the plains of northern India (Punjab) showed higher frequencies than yellow *A. mellifera* (presumably *A. m. ligustica*) from Europe: 306 wing beats/s vs. 235 for workers and 283 vs. 225 for drones (Goyal and Atwal 1977). These frequency differences could well be a mere consequence of wing size differences. Wing beat frequencies of *mellifera* bees of comparable size have not yet been determined.

9.4.1.6 Division of Labor

The characteristic sequence of activities in the life history of a worker bee, well established by many publications in *A.mellifera*, has hardly been investigated in *A.cerana*. Unpublished investigations carried out in the flight room of the Apicultural Institute in Oberursel (thesis of J.Perk 1973) with a *cerana* colony from Pakistan showed that the same sequence of functions exists as in *A.mellifera*, each function starting, however, a little earlier: cell cleaning before the third day; brood care, pressing the pollen loads in the cells and comb construction beginning with the third day. The maximal development of pharyngeal glands was observed between the 4th and 16th day; the function of wax glands started on the 3rd day, with the maximum development occurring between the 12th–16th day and regressing from the 22nd day. The first orientation flights were observed between the 7th and 11th day.

9.4.2 The Nest

9.4.2.1 Nesting Site

Feral colonies of *A.cerana* are found in similar locations as *A.mellifera* colonies – hollow trees, clefts in rocks and walls etc. (Fig.9.3; Fan Tsung-De 1956; Morse and Benton 1967; Tokuda 1971; Schneider and Kloft 1971). The traditional hives constructed for this bee are substantially smaller than those used for *mellifera* (Fig.9.32; Okada et al.1958; N.Koeniger 1979). Six to eight combs were counted in feral colonies in India (Muttoo 1956), 4–10 in Ussuria (Lavrekhin 1958), 5–9 in Japan (Okada 1985). In Thailand, Seeley et al. (1982) counted 5.6 combs on the average with an area of 2825 cm^2 (one side; n = 15). The number of bees per colony was 6884 $\pm$ 3418 (n = 3). Attempts to keep *A.cerana* colonies in standard size Langstroth hives have failed in India (Pandey 1977) and Thailand (Akratanakal 1976) as well as in Japan (Sakagami and Kouta 1958). In coconut plantations in Malaysia, where *cerana* colonies thrive exceedingly well, they nest in heaps of husks or discarded broken shells, hollows of tree trunks and rafters of houses. Lacking a cave of sufficient size, colonies are usually small (1400–2000 bees; Makhdzir and Osman 1980). In Japan, a *cerana* colony consists "usually" of 10–20,000 bees (Okada 1985).

9.4.2.2 Combs

The wax of *A.cerana* has a melting point (65°) about 2° higher than that of *A.mellifera* (Padhke 1961, Tokuda 1971). The beeswax exported from India, commercially known as Ghedda wax, comes probably mainly from *A.dorsata* and to a

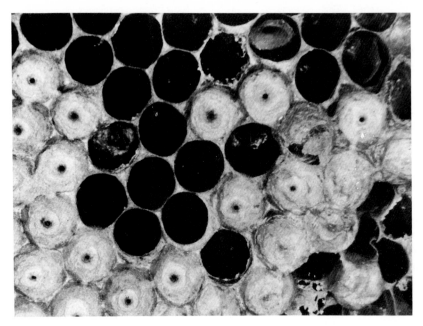

Fig. 9.13 Capped drone cells of *A. cerana* with pores

small extent from *A. cerana*. Cell diameter in *A. cerana* is relatively small compared to body size: For worker cells 4.78 (Japan, Okada and Sakai 1960), 4.6 mm (Ussuria; Lavrekhin 1958) to 4.37 mm and 4.25 mm (plains of India, Muttoo 1956; Deodikar et al. 1958); The measurements by the author showed an even more important variation: in the north 4.87 mm (Peshawar) and 4.67 mm (Peking), in the south 4.25 (Thailand) and 4.20 (Sri Lanka). The diameter of drone cells was determined to be 7.08 mm (Japan). The difference in size between worker and drone cells is less pronounced in *A. cerana* than in *A. mellifera*: Ratio 1:1.13 vs. 1:1.33 (calculated from data of Okada and Sakai 1960).

Dismantling old combs seems to be a specific trait of *A. cerana*, resembling the way in which *A. florea* treats its absconded nest (Chap. 7). While *A. mellifera* reuses the combs as long as the colony lives, *A. cerana* gnaws the wax structure of dark combs down to the middle sheet in order to build new cells on it (Kellog 1941; Atwal and Dhaliwal 1969; Tokuda 1971; Okada 1985). The wax debris falls to the bottom of the hive and accumulates there, the cleaning behavior being poor, providing an excellent medium for wax moths. A seemingly "hygienic behavior" therefore changes into the opposite.

Another characteristic of the *cerana* comb, a pore in the drone cocoon, is evidently found in the whole area of distribution (Jacobson in Buttel-Reepen 1907; Roepke 1930 - Java; Tokuda 1922 - Japan; Kellog 1941 - China; Lavrekhin 1958 - Siberia; Lindauer 1956 - Sri Lanka; Hänel and Ruttner 1985 - Pakistan and Malaysia). As in all honeybees, the drone cell is capped by the worker bees with a wax cover and then the drone larva spins the cocoon. However, 1–3 days after completion of the cover the bees already start to remove the wax and a yellowish,

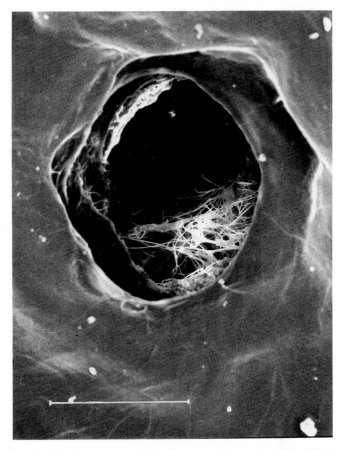

Fig. 9.14 Tunnel of pore from inside: homogenous "melted" mass and rest of cocoon. SEM. Horizontal line 0.2 mm. (Hänel and Ruttner 1985)

hard, silky plate appears. At its center lies the pore (Fig. 9.13; Sakagami 1960). It is funnel-like, measuring 0.4 mm in depth and 0.25×0.5 mm in diameter, the outside opening being broader than the inside, and it is fringed at the inside edge with a heavy brown substance (Fig. 9.14). This pore is the result of the local dissolution of the cocoon from inside (presumably by the action of a drone larva secretion, a special cocoonase) while the wax cover is still intact (Fig. 9.15, Hänel and Ruttner 1985) The real significance of this unique structure is unknown.

Wherever the behavior of *A. cerana* has been studied, no use of propolis was found. It was frequently observed that cracks in the hive are not sealed. However, a rather brittle, light greyish mass is present on hive walls and frames, which evidently is not pure wax.

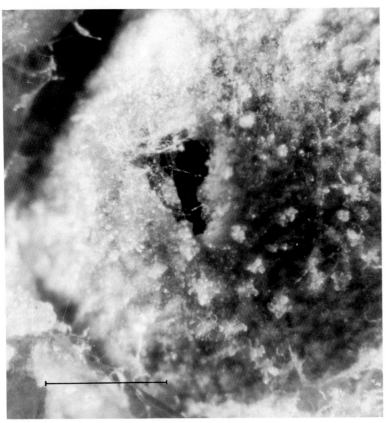

Fig. 9.15 Drone cell with burst cupola. Solitary cocoon filaments crossing the fissure are visible. Horizontal line: 0.2 mm. (Photo Hänel)

9.4.2.3 Colony Defense

A. cerana is generally reported as being "mild, tolerant and timid" in defense behavior. These attributes are certainly imputed to this species on account of weak resistance to the attacks of its co-species and new competitor, *A. mellifera*, which was only recently introduced, and to its reactions to manipulations by humans. However, *A. cerana* shows a number of behavioral patterns which prove to be very effective against traditional enemies.

a) Abdomen shaking (Sakagami 1960; "body shaking" Koeniger and Fuchs 1975): a collective violent lateral shaking of the slightly raised abdomen is released by approaching insects or by any moving dark object visible against a bright background. The repellent effect of this behavior on insects is clearly visible (Butler 1954, Koeniger and Fuchs 1973).

b) Hissing behavior (Butler 1954; Schneider and Kloft 1971; N. Koeniger and Fuchs 1973): short knocks on the beehive or gentle blowing on the combs or in a swarm induce a sharp hissing sound which lasts about 0.5 s. The sound is pro-

Fig. 9.16 "Group defense behavior" of *cerana* workers near the nest entrance in a rock. (Photo P. Schneider)

duced by a collective quick movement of the wings (closing over the body, while the abdomen jerks upward). The reaction is transmitted to bees, besides immediate contact with the stimulating agent, by body contact; it migrates over the comb or the bee cluster with a velocity of 3 cm per s. Airborne sound induces hissing behavior only if it is transmitted by substrate vibrations, that is to a cluster of bees sitting on a comb but not to one on a block of cement. The hissing sound (which resembles the sound produced by snakes) seems to have a repellent effect on larger vertebrates (Koeniger and Fuchs 1975).

c) Group defense (Schneider and Kloft 1971): if attacked by powerful enemies such as hornets, *cerana* bees do not counter-attack, as most races of *mellifera* do (which in this case gives the defender only a small chance), but, with the tip of the abdomen raised, they form groups of 30 near the entrance (Fig.9.16). The "timid" *cerana* abruptly stops flight activities on the arrival of a hornet. The hissing sound described above is repeatedly emitted in the hive. After this rapid retreat with no solitary counter-attack the hornet usually relinquishes its attempt and leaves.

The dramatic events which are observed during an attack of the Japanese Giant hornet, *Vespa mandarinia*, were described by Matsuura and Sakagami (1973) in a fascinating report. If hornets persist in the attack, they do not dare to capture bees out of the group (Fig.9.17). No bee takes to the wing. If the hornet approaches too closely, it is seized simultaneously at the legs and wings by several bees and "drowned" in the mass of bees and killed. In most of the observations, virtually no bees died during the defense. The hornet attacks are stopped at the

135

Fig. 9.17 *Vespa mandarinia* attacks a *cerana* colony. Initial phase: the bees stop flying, raise the abdomen and expose the scent gland. (Photo F.Sh.Sakagami)

very beginning. Morse and Benton (1967) observed a similar group reaction while using their bee venom collection device at the entrance of a *cerana* hive: the whole colony moved out and covered the apparatus densely.

The attacks on *mellifera* colonies by *Vespa mandarinia* in Japan develop quite differently: at a hive of *A. mellifera ligustica*, *mandarinia* wasps wait for the attacking bees, seize and kill them with their strong mandibles and transport the corpses back to the nest (Fig.9.18; "hunting phase"). When the number of hunting wasps increases and the defense reactions of the bee colony become weaker, the "slaughter phase" starts. The wasps kill one bee after the other without carrying the corpses to their nest, until the resistance ceases and they succeed in occupying the beehive (phase 3) and eating or transporting pupae and larvae. At the entrance of the occupied hive, territorial defense behavior of the wasps is observed (Matsuura and Sakagami 1973).

In Japan, *A. cerana* is the only potential prey of *V. mandarinia* which has developed an effective defense tactic. Even colonies of other species of large wasps – *V. crabro, tropica, mongolica, analis* – are attacked and in some cases eventually slaughtered. In India, *A. cerana* shows the same behavior when attacked by large wasps (*Vespa magnifica, V. orientalis*; Sharma et al. 1980). Group defense behavior evidently represents a higher level of cooperation in social nest defense than that of individual counter-attacks.

According to reports by Japanese beekeepers only *A. mellifera* is attacked by

Fig. 9.18 *Vespa mandarinia* catches a *mellifera* worker attacking on the wing. (Photo M. Ono)

V. mandarinia if colonies of both species are kept at the same location. These wasps are a serious threat to beekeeping in Japan, and *mellifera* colonies have to be protected by a grid at the entrance. To robbing attempts by *A. mellifera* no effective defense reactions are developed in *A. cerana*. Frequently there are no guard bees at the entrance, intruders may pass without being inspected and sometimes bees of a robbed *cerana* colony were observed to feed the robber bees. During the hot season a *cerana* colony attacked by *mellifera* bees usually loses all its stores and then absconds or dies. "Chinese bees rob Italians in the winter (see p. 138) and in turn they are robbed by the Italians in the summer. When Italian bees come into a neighborhood, native bees disappear" (Kellog 1972). This is what happened to a large extent in Taiwan and Japan (Akahira and Sakagami 1958) and also in large areas of the PR of China (Ma De-Feng and Huang Wen-Cheng 1981).

d) Stinging behavior: *A. cerana* is in general less prone to sting than *A. mellifera*. The amount of isopentyl acetate, the alerting pheromone, in the sting is about half the quantity found in a *A. m. ligustica* worker (~ 0.8 and $\sim 1.9\mu g$, resp.; Morse et al. 1967). A target of black felt ("artificial mouse") with two drops of iso-amyl acetate, used to test the defense reaction at the hive entrance of a *cerana* colony in India, was soon covered by attacking bees. Afterwards, however, no stings were found in the felt (Morse and Benton 1967). In a similar experiment with *mellifera* the felt would look almost white because of the stings left in it. Several highly developed characteristics which result in sting autotomy in *A. mellifera*, are clearly reduced in *A. cerana* (Sakagami 1960b).

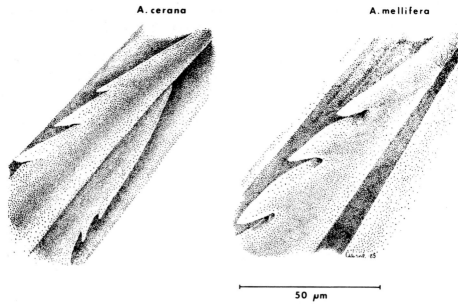

A. cerana   **A. mellifera**

50 μm

Fig. 9.19 Sting lancet of *A. cerana* and *A. mellifera* (Design A. Aarhuis)

1. Frequency of sting apparatus autotomy is reduced in *A. cerana* by a change in behavior, that is by revolving movements after stinging (instead of straight runs) and by strong muscles anchoring the sting apparatus to the spiracle plates.
2. Reduction of the sting protracting muscles. Similar to Meliponinae, *A. cerana* uses its mandibles instead of the sting when attacking an enemy. Furnished with remarkably strong muscles, they are used for gnawing while persistently sticking to an intruder.
3. Of all *Apis* species, *A. cerana* has the least developed barbs on the lancets of the sting (Weiss 1978). In specimens from Sri Lanka, the first barbs were rudimentary, the others shorter and more depressed (smaller angle of the barb to the lancet axis; Fig. 9.19).

The venom of *A. cerana* is identical with that of *A. mellifera* in the amino acid sequence of the melittin, its main component (Kreil 1973). Alarm substances: Isopentyl acetate was found in worker bees from the Philippines and Thailand, but in much lower quantities than in *mellifera* (about 1 μg/bee vs. about 2 μg/bee). However, the two strains examined showed a great quantitative difference (Morse et al. 1969).

9.4.3 Field Activities

9.4.3.1 Temperature Dependence

A. cerana is reported to be active at lower temperatures than *A. mellifera*. Field activities have been observed in Ussuria at 6-8° (Danilova 1960). The same was noticed in Kashmir (Shah 1980) and China (Oschmann 1961, Gong 1983). In

Himachal Pradesh *mellifera* is the more active species during the warm season, *cerana* during the cold one (Adlakha and Sharma 1974). *Cerana* collects nectar from *Plecthanthus* blossoms on cool autumn days, while *mellifera* does not (Sharma et al. 1980). In Japan *cerana* starts flying in spring 1 month earlier than *mellifera* (Miyamoto 1958). Therefore relations between the two species are sometimes reversed. During cool morning hours, while *mellifera* remains still inactive, *cerana* is robbing their nests (Goyal 1974). Again, these behavioral differences might not be species-linked, but could be specific adaptations of certain ecotypes (see the experimental data on isolated bees, p. 127)

9.4.3.2 Flight Pattern

The flight of *A. cerana* resembles that of a fly – rapid, hasty, unpredictably zig-zag – compared to the steady, clumsy flight of European *mellifera* bees (Kellog 1941; Lindauer 1956; Sakagami 1960). This behavior helps in escaping from flying predators, like hornets and bee-eating birds. Two *mellifera* colonies, transported to Sri Lanka, immediately attracted several Merops birds and lost all their field bees, in spite of protecting measures, while indigenous *cerena* and *florea* colonies continued their habitual activities (N. Koeniger, pers. commun.). It has to be investigated, however, whether this flight pattern is species-specific and whether some tropical *mellifera* races show the same behavior. When bees were dislocated 50 m, the speed of homing was found to be shorter for *A. cerana indica* (192 s) than for *A. m. ligustica* (295 s; Atwal and Dhaliwal 1969).

9.4.3.3 General Flight Activity and Flight Range

In a flight room a *cerana* colony from Poona (India) with permanent light and feeding within the hive showed 5.5 times as much flight activity relative to the number of bees of a colony of *A. mellifera carnica* (Institute Oberursel, unpubl. thesis 1971 by E. Preuhs).

Lindauer (1956) was not able to train *cerana* workers in Sri Lanka farther than 750 m from the hive. During his experiments the flight distance of bees collecting on natural sources did not exceed 300 m. A main foraging distance on cauliflower (75% of the bees) of up to 400 m, on barberry 600–700 m, was found in Punjab, India (max. distance 900 m, or 1100; Dhaliwal and Sharma 1973). The uphill foraging distance was much shorter (300 m) than in flat country (Dhaliwal and Sharma 1974). A nectar flow 1000 m away was not visited by the bees of a village. In consequence, migratory beekeeping became indispensable (Chonduri 1940). Few data are available so far for the flight range of the temperate zone *A. cerana*. In Kashmir, *cerana* bees were observed to collect pollen from *Crocus sativus* at a distance of 3.75 km (Shah and Shah 1980b). A number of observations are available regarding different flower preferences between the two species. *A. cerana* (from Japan) is reported to visit a greater variety of plants, including wild species, while *mellifera* foraged mainly on *Trifolium* and *Brassica* (Miyamoto 1958). In

Himachal Pradesh the main nectar source of *A.cerana, Plecthranthus* sp., is neglected by *mellifera* foragers (Goyal 1974; Ahmad 1982).

Honey collected from *cerana* bees (originating from China) in Germany, showed differences in the pollen spectrum compared to *mellifera* honey collected simultaneously at the same location (Vorwohl 1968).

9.4.4 Orientation and Communication

In principle *A.cerana* shows the same behavioral characteristics during foraging activities as *A.mellifera*: sun orientation (even on cloudy days), direction transposition to the vertical plane, round and waggle dances with sound production (Lindauer 1956). Sound production during the waggle run very closely resembles that of *A.mellifera*. The fundamental frequency and burst rate of the *A.mellifera* and *A.cerana* sounds are roughly the same for both species (290 Hz and 30 bursts/s respectively; Gould et al. 1985). This, too, shows the close relationship of the two species.

In detail, however, clear differences are found. The dance rhythm for each distance indication is slower in *cerana* than in *mellifera,* the decrease in number of waggle runs is steeper for close distances (up to 150 m), slower for greater ones (Fig. 4.9). This implies that distance indication by the bees becomes more precise for food sources in close vicinity and less so farther away (see flight range). In dance runs *cerana* has the tendency to repreat two or more runs in one and the same direction before turning around and revolving in the opposite direction. The transition from round dance to waggle dance occurs at very close distances (0.5 m), Lindauer 1956; 7–8 m, Goyal and Atwal 1971). Frequently the round dance contains an indication of direction: turning points during the round runs occur at the position which corresponds to the direction to the food source (Fig. 9.20). This intermediate type was described as "directed round dance" or "modified sickle dance" (Lindauer 1956).

More training acts were needed for *cerana* in India to remember a feeding situation than for *mellifera* (Goyal and Atwal 1971).

Menzel et al (1973), however, working in Germany with *cerana* from Pakistan, observed the contrary: *cerana* was the faster in learning a given situation. Koltermann (1973a, b) observed the same degree of retroactive memory inhibition in both species if two food signals of different quality were given successively. The two species (and three races of *mellifera*) showed different spontaneous scent preferences. In *A.cerana* the following sequence of scent attractivity was found: rosewood oil – lavender – rosemary – fennel – cinnamon alcohol (15 scents tested). In *A.m.ligustica* the sequence was: pine needle extract – lavender – camomile – rosewood oil.

9.4.5 Processing the Nectar in the Hive

Of the enzymes added by the bees to the nectar as they transform it into honey, the level of invertase is about the same in *A.cerana* from China as in *A.mellifera* (Vorwohl 1968). In *cerana* honey collected in India a higher invertase level was

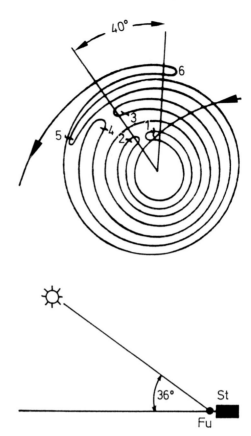

Fig. 9.20 Communication of sun orientation by directed round dance in *A. cerana*. (Lindauer 1956)

observed (Wahlke and Desai 1980). As far as the amylase activity is concerned, less than half of the quantity present in *mellifera* honeys was found in a study on Chinese *cerana* bees in Europe as well as in commercial honey originating from China (Vorwohl 1968). In fresh honeys from India, however, Walkhe and Desai (1980) reported diastase values similar to those known for *mellifera* honeys. Further investigations are needed to achieve generalized statements.

A major step in producing a durable honey is to reduce the moisture content of the nectar to 17–20%. During a nectar flow considerable amounts of water have to be removed from the hive. Ventilation by the bees near the entrance is one of the activities to achieve this. *A. mellifera* workers fan with their heads towards the entrance, *cerana* in the opposite direction, the head turned away from the entrance (Fig. 9.21). This very obvious behavioral characteristic was noticed by practically all observers throughout the *cerana* area. It is not influenced by the environment. In mixed colonies, with workers of both species fanning side by side, each keeps its inherited position (Dhaliwal and Atwal 1970).

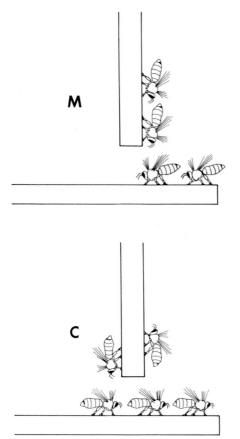

Fig. 9.21 *C* Fanning position of *cerana* workers at the hive entrance: head directed away from entrance, *M A. mellifera.* (Sakagami 1960 a)

9.4.6 Interspecific Relations

Koeniger and Vorwohl (1979) studied the interactions between sympatric bee species in Sri Lanka, including *Trigona viridipennis*, at an artificial feeding dish. Smaller bees behaved with less tolerance towards larger ones than vice versa. That is, *A. cerana* was attacked by *Trigona* and *A. florea*, but it in turn attacked *A. dorsata*. Because the plants exploited were not the same, honey produced by *A. cerana* was different from honey of other species.

In a common feeding location, *A. mellifera* was clearly dominant over *A. cerana*, and also more aggressive. It soon excluded bees of the other species from the feeding dish (Sakagami 1959; Dhaliwal and Atwal 1970).

A number of experiments were made with mixed colonies of *cerana* and *mellifera* workers (Sakagami 1960; Dhaliwal and Atwal 1970; Ruttner unpubl. data). Only a limited coexistence of bees of the two species in the same colony was achieved. Brood of the other species is removed as far as eggs and larvae are concerned. Adult bees, hatched from heterospecific sealed combs, are accepted and they participate in social work, at least partially. In certain cases, however, *cerana* bees were expelled by *mellifera* workers (in Japan, Sakagami 1960) or homing *cerana* field bees were attacked by *mellifera* guard bees (*cerana* from China; Ruttner unpubl. data).

Adult bees introduced into a colony of the other species were fairly well accepted in both directions as long as they were very young (less than 1 day old). Up to 85% *cerana* bees, 3 days old, were expelled from *mellifera* colonies within 3 days, but in the reverse situation (*mellifera* in *cerana* colonies) not more than 18% were driven away. Bees older than this were accepted in both directions only in very low numbers. Successful introduction of mated *mellifera* queens into *cerana* colonies was reported (Atwal 1968; Adlakha and Sharma 1971; Goyal and Atwal 1979; Sharma et al. 1980). If mature queen cells were inserted, *mellifera* queens hatched among *cerana* workers, they were mated and started egg laying (Adlakha and Sharma 1971). Rearing *cerana* larvae in *mellifera* colonies to queens was tried with little or no success (Tokuda 1971; Akratanakul 1976; Ruttner unpubl. data). The reason for the failure was evidently the composition of the royal jelly in respect to its effect on metamorphosis. As double graft, eight *cerana* larvae were accepted in a queenless *mellifera* colony. Four cells were sealed; however, no queen hatched. Opened after the due time for hatching, each cell contained a very big larva without any sign of pupation. Inoue (1962) grafted 100 larvae of *A. cerana japonica* into *A. mellifera* colonies, and obtained only three queens, which were not accepted by the *A. mellifera* colonies from where they emerged.

9.5 Reproduction

9.5.1 Daily Egg Rates of the Queen

In *cerana* colonies relatively low figures are reported for the maximum number of eggs deposited daily: 300-500 in the south of India (Muttoo 1956), 700 and 830 in Kashmir and northern India (Kapil 1957; Sharma 1960a; Saraf and Wali 1972). Only Shah and Shah (1980b) present figures from Kashmir which correspond approximately to those found in *A. mellifera* in other countries (1440-2030 eggs per day).

Atwal and Sharma (1970) compared the brood areas of *A. cerana indica* (nine colonies) with those of different *mellifera* strains (21-24 colonies) in Nagrota, northern India. The former produced only 1/4 to 3/4 as much brood as *ligustica* colonies. These figures, however, do not nessarily represent the egg-laying capacity of the queens, since they were not based on egg counts, but on conclusions drawn from the number of sealed cells. As was shown by Woyke (1976a), this method will not always give reliable results. While about 95% of the larvae were sealed during good nectar flow (in Poona, Maharastra, South India), only 50% could be observed as sealed, however, when nectar and pollen intake was moderate. During a shortage of food supply the queens continued to lay, but the worker bees ate all the larvae. Thus it is not necessarily the low egg-laying rate of the queen, but the variable and sometimes low level of brood-rearing efficiency which results in a scattered broodnest and low brood production of *cerana* colonies.

Reddy (1980) noted that mean brood areas fluctuated somewhat above 2000 cm^2 during months of highest brood production (March-October) in South India. Only once in a 2-year period (July 1975-June 1977) did this figure exceed 3000.

The number of ovarioles in *cerana* queens is reported to be less than half ("about 73"; Kapil 1962) of that in *mellifera* queens. However, it is a question whether this anatomical difference alone could explain a lower egg-laying capacity.

9.5.2 Laying Workers

These appear within a few days after dequeening (Tokuda 1924; Akahira and Sakagami 1958). In central China laying workers were reported to appear 2-3 days after dequeening, filling up to six combs with eggs (Blandford 1923). In spite of this, queen rearing is practiced in China by the standard methods (Oschmann 1961). For grafting larvae, only cups with a small diameter (up to 8 mm) are accepted. The number of ovarioles in worker bees is about double that generally found in *mellifera* bees of European race (7-9, Alpatov 1938a; Sakagami and Akahira 1958). Kapil (1962) reported a mean number of 4.58 ovarioles of *cerana* bees of north India per ovary.

Ovaries of worker bees 10 days old are much larger than in *mellifera* races except *capensis* (Fig. 9.22). During early summer 10-20% of the workers showed developed ovaries with mature eggs in queenright colonies (Sakagami and Akahira 1958).

9.5.3 Swarming

9.5.3.1 Reproductive Swarms

Swarming connected with the rearing of young queens and fission of the colony is a regular event in the seasonal cycle of *A. cerana*. Preparations for swarming start as soon as the combs are filled with brood, pollen and honey. In one experiment performed in Pakistan to study different management methods, Latif et al. (1960) found swarming tendencies in 100% of the normally kept, single-queen colonies. In northern Pakistan, swarming will start as soon as a colony reaches a population of around 20,000 bees. An average of eight swarms per colony was reported there (N. Koeniger 1976a). In Japan, however, only one swarm ("or two to three") is observed annually per colony (Tokuda 1971).

Queen cells are built in the lower margin of the central combs (or also on the upper part; Japan, Sakagami 1960; Okada et al. 1958). They soon become inconspicuous because of an intensive construction of worker cells around them (Fig. 9.23). Emergency queen cells, constructed by transforming worker cells, are also hard to see, being depressed on the surface of the comb. During artificial queen rearing (grafting of larvae), however, the cells remain free and isolated until hatching. The shape of the queen cells is broad and short (surprisingly short compared to the size of the queen, Fig. 9.24). At the end of the pupal period the wax is

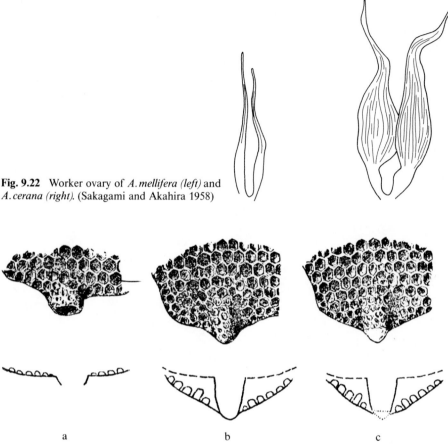

Fig. 9.22 Worker ovary of *A. mellifera (left)* and *A. cerana (right)*. (Sakagami and Akahira 1958)

a   b   c

Fig. 9.23 a-c Development of a queen cell (**a-c**). (Sakagami and Kouta 1958)

removed from the tip, and the brownish cocoon appears. The capping of the cell is cut open by the queen with her mandibles.

Apicultural experience indicates that the time of development of *cerana* queens is about the same as in *mellifera*. In one single recorded experiment the queen hatched on the 16th day (Shah and Shah 1980b).

The number of queen cells in 52 colonies under swarming impulse varied between 2 and 29 (average 9.0). These observations were made in Katrain, Punjab, in the mountains of northern India, at an altitude of 1700 m (P. L. Sharma 1960). The majority of the swarms issued before 13.00 h. The maximum weight of the swarms was 1.8 kg (= 16,000 bees), average 1.0 kg, corresponding to the generally small size of colonies. The number of bees in the afterswarms was sometimes only 2000. *Cerana* queens (from Pakistan) are about 12% smaller than *carnica* queens. However, on account of the broad vaginal passage, injection of semen into the median oviduct during instrumental insemination succeeds easily (Ruttner et al. 1973; Woyke 1973).

Fig. 9.24 *Cerana* queen cells (oft-grafted queen cups) after hatching

In Japan *cerana* swarms choose a rather broad substratum (e. g., underneath an inclined trunk) when forming the initial cluster; it is attached to it like a flat cake (in contrast to the cluster-shaped swarm of *mellifera*; Sakagami 1960).

9.5.3.2 Migratory Swarms

Seasonal migration of bee colonies is a common characteristic of all (or most) tropical honeybees. It has not yet been investigated whether it occurs in all geographic types of *A. cerana*; this is not very likely, since in many reports on bee-keeping with *cerana* it is not mentioned. In Taiwan regular seasonal migration of *cerana* colonies occured between the plains and the humid mountain forests (Fen Tsung-Deh 1952).

9.5.3.3 Absconding

During absconding the colony completely deserts the nest, just as in migratory swarms, the difference being that this does not necessarily occur seasonally and that the flight direction is not predetermined. There may still be eggs in the combs of the deserted nest (Woyke 1976a). Sakagami (1960) enumerates several causes for this escape of a whole colony: (1) propagation of wax moths, (2) food shortage, (3) weakening by robbery or absence of a queen, (4) unsuitable environmental conditions because of high temperature or lack of ventilation, (5) disturbance by beekeepers or by enemies. The colonies often abandon their home also without any apparent external causes. According to Sakagami (1960) absconding occurs also in the temperate zone of Japan, but evidently less frequently than in S Asia. *Cerana* colonies of N China, kept for several years in Central Europe, never showed any tendency to abscond, in spite of frequent disturbance (Ruttner unpubl. data). In Kashmir, absconding is less likely than in other parts of N India (Verma 1984). In Thailand, however, 100% of the colonies abscond after the honey harvest. (Akranatakul 1984). Absconding is prevented easily by transporting colonies to areas with sufficient nectar flow (N. Koeniger 1976).

9.5.4 Mating Period and Start of Egg Laying

9.5.4.1 Mating Flights

a) Drones. Evidently, timing of mating flights is an important factor for establishing isolation between different *Apis* species. In Sri Lanka, Koeniger and Wijayagunasekera (1976) found three different daily drone flight periods for the three sympatric species. The flight of *cerana* drones was restricted to 1 h, 16.15–17.15 h

However, in Europe the flight time of drones of *cerana* colonies imported from Pakistan was about the same as that of local drones: 11.15–15.15 h, maximum between 12.15 and 14.15 h. The activity of drones from Peking lasted one hour longer. In Japan, mating flights of *A. mellifera* and *A. cerana* occur during the same time of day (Hoshiba et al. 1981). It was observed that *cerana* drones visited a well-known *mellifera* drone congregation area near Frankfurt, Germany. Out of 2599 drones caught on 3 consecutive days while circling over this area, 53 (= 2.12%) were *cerana* drones. They came from a population of about 1000 drones in two *cerana* colonies 500 m away (Ruttner 1973b). The frequency at the congregation area corresponded very well with their overall frequency in this location. This observation is a remarkable indication that drone orientation during flight follows a super-species pattern.

b) Queen. The behavior of young queens before, during, and after mating is quite similar to that of *A. mellifera* (Shah and Shah 1980). However, in northern India (Kashmir, Punjab) queens started flying at an earlier age than is reported for *A. mellifera* queens (review see G. Koeniger 1986): orientation flight at 3–4 days, mating at 4–6 days (P. L. Sharma 1960a; Adlakha 1971; Shah and Shah 1980).

147

It has to be considered whether and how much the mating behavior is influenced by environment. *Cerana* queens in Europe mated at an older age (though the experimental methods may have biased the results; Ruttner et al. 1972). On the other hand, *mellifera* queens in India (Kangra Valley) started to fly and mated earlier (4 and 5 days old respectively; Adlakha 1971) than in zones with temperate climate. P.L. Sharma (1960a) and Adlakha (1971) give similar figures on the duration of the *cerana* queen's mating flight: orientation flight 1–8 min, mating flight 15–29 min. In S India flights of queens were observed between 13.30 and 15.30 h, and mating flights between 14.00 and 15.00 h with an average duration of 27 min (Woyke 1975b).

Cerana queens produce 9-oxodec-trans-2-enoic acid (9-ODA), like queens of the other three *Apis* species (Butler et al. 1967; Shearer et al. 1976). Free- flying *carnica* drones are attracted to a tethered *cerana* queen nearly as much as to a queen of their own race. The stimulation of the receptors (pore plates at the antennae of drones) by secretions of the mandibular glands of queens of both species, or by synthetic 9-ODA, in electrophysiological experiments was the same in *A. cerana* as in *A. mellifera* (Ruttner and Kaissling 1968).

In the country of origin, the mating success is evidently high. In an experiment with 310 queens during 3 years, only 44 (= 14%) were lost (Shah and Shah 1980). *Cerana* queens in a *mellifera* area, however, are not successfully mated with conspecific drones even if these are present in sufficient number. No matings at all were achieved when *mellifera* colonies were located at a distance of a few hundred meters only. Mating success (rate of laying queens, number of spermatozoa in the spermatheca) increased with increasing distance from *mellifera* colonies (Ruttner and Maul 1983). Similar difficulties are reported in reciprocal conditions (*mellifera* queens in a *cerana* area; Sharma 1960; Akratanakul 1976; Ahmad 1984; Cadapan 1984). Only as soon as an important *mellifera* population was finally established (as in Japan, China, and some parts of northern India) was the usual mating success achieved again. Ahmad (1982) observed good mating success with *mellifera* queens in Islamabad and Hydarabad by promoting very early raising of drones and queens.

Having the same flight time and the same reaction to the sex attractant at the same congregation areas, drones of the other species evidently interfer with conspecific matings. It was shown that *mellifera* drones actually mate with *cerana* queens, though with noxious effect on the queen: a young *cerana* queen was found with its damaged sting chamber firmly blocked by the mating sign of a *mellifera* drone (Ruttner and Maul 1983). It can be concluded from all these observations that no premating barrier exists between these two species as is the case between the other species. Thus no permanent coexistence of the two species seems to be possible.

Because of these data of a not completely finished speciation it appears totally unjustified to classify this taxon (*Apis cerana*) as subgenus as was proposed by Skorikov (1929b) and Maa (1953); (see Chap. 4). On the contrary, they have to be regarded as being in a late, but not yet finished stage of speciation.

9.5.5 Mating

The effectuated copula is indicated by a "mating sign" in the sting chamber of the queen returning from a flight (Fig. 9.25). The *cerana* mating sign is an elongated, whitish plug of mucus, thickly covered by a reddish, shining substance, supposedly the coat of the cornua of the endophallus.

In *A. mellifera*, about half of the queens return twice to the hive with a mating sign. Two or more mating flights of the same queen is an indication of multiple mating (on the average 8–10 matings per queen, frequently during one single flight) which is believed to be obligatory to transfer sufficient spermatozoa for the queen's lifetime (Ruttner 1985; G. Koeniger 1986). In *A. cerana*, P. L. Sharma (1960a) observed two out of nine *cerana* queens returning twice on two consecutive days with a mating sign. In an experiment performed in Germany with *cerana* colonies from Pakistan, three queens out of eight were observed to have two effective mating flights (two returns with mating sign; Ruttner et al. 1972). Thus the number of effective mating flights might be the same in both species, though Adlakha (1971) reported only single mating flights in 12 observed queens. However, a number of pre- and postmating flights, without mating sign, was observed by all authors, though less than in *A. mellifera*.

One important difference between the two species is the number of spermatozoa counted in one drone: 10–11 mill in *mellifera*, 1 mill in *cerana* (Woyke 1975b). Yet the spermatheca of a naturally mated *cerana* queen contains only about 30% less spermatozoa (3.5 mill) compared to *mellifera* (Ruttner and Maul 1983). In India, Woyke (1975b) even found only 1.3 mill spermatozoa within the spermatheca of young laying queens. It was calculated that a *cerana* queen has to mate

Fig. 9.25 *Cerana* queen with mating sign. (Photo Eidam)

with up to 30 drones to achieve a completely filled spermatheca. One queen was dissected after the return of the second effective mating flight; the spermatheca was filled with a not determinable number of spermatozoa (from the matings of the flight before) and the oviducts contained 2.8 µl of semen, corresponding to the amount produced by 14 drones.

The process of transferring the spermatozoa from the oviducts into the spermatheca is not more efficient in *cerana* than in *mellifera*. Experiments with cross-insemination show rather the contrary (Woyke 1973). Thus the lower quantity of semen (explained anatomically by smaller testes and vesicles; Ruttner et al. 1973) can only be compensated for by a higher number of matings. It is surprising that in spite of this, on the average relatively short flights were reported: orientation flights 1–8 min, mating flights 15–29 min.

Instrumental insemination of *A. cerana* queens meets with problems mainly because of the drones. About 40 drones are needed to collect the semen for one queen. Moreover, the high irritability of *cerana* drones causes additional difficulties (Woyke 1973).

9.5.6 Start of Oviposition

Cerana queens frequently started laying eggs 2 days after mating. At that time they were generally only 8 days old (Shah and Shah 1980).

9.6 Genetics

9.6.1 Genetic Incompatibility of the Cross mellifera x cerana

Heterospecific instrumental insemination of *A. mellifera* and *A. cerana* (with semen of the reciprocal species) is feasible: spermatozoa enter the spermatheca, they are able to survive there and to descend again through the spermathecal duct to fertilize an egg. During the first 24 h after fertilization an undisturbed cleavage is observed to the blastula stage of the zygote. Then, however, cell walls start to disintegrate and nuclei to migrate into the secondary periplasm to accumulate in the antero-ventral part of the zygote and to degenerate completely later on (Fig. 9.26).

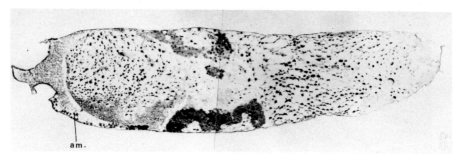

Fig. 9.26 Hybrid embryo *mellifera × cerana* with defective development. *am* amnion (Photo Maul)

No hybrid larva or imago ever developed in these experiments (Ruttner and Maul 1983).

No morphological differences were found between the karyograms of the two species (Hoshiba et al. 1982). The amino acid sequence of melittin, the major compound of bee venom, was found to be identical in the two species *mellifera* and *cerana*, but not in *dorsata* and *florea* (Kreil 1973, 1975, Fig. 3.4).

Contrary to this result and to the overall similarity between *mellifera* and *cerana* are the data on DNA hybridization experiments published by Sperlin et al. (1975). These authors found that about 60% of the *A. melifera* genome *(ligustica, anatoliaca)* can form heterohybrid DNA with *A. cerana* (from India), while 40% cannot. This is an extraordinarily high level of divergence, considering that single-copy cow DNA formed 93% stable heterohybrids with sheep DNA. Based on these data, the authors speculate on a possible exceptional speed of evolution in these species. However, how can a high degree of dissimilarity of DNA and the evident phenetic similarity be brought into agreement? Therefore a confirmation of the data of DNA hybridization experiments seems to be indispensable.

9.6.2 Sex Determination

Sex determination is evidently the same in *A. cerana* as in *A. mellifera* (Woyke 1979; Hoshiba et al. 1981): the diploid females are heterozygot in the sex locus ($x_1 x_2$), haploid males hemizygot (x_1). By brother-sister mating diploid homozygotic individuals ($x_1 x_1$) originate, that normally are eliminated in the early larval stage, but that can be raised artificially to adult diploid drones. Woyke (1980) reported that diploid drone larvae of *A. cerana* (from India) were eaten by workers at a later stage than were *mellifera* larvae, and that some of the diploid drones reached the adult stage within the colony.

9.7 Geographic Variability

9.7.1 History of Systematics

The taxonomic analysis within the species indicates a similar historical evolution as shown in *A. mellifera* (Chap. 5). Early taxonomic classification in *A. cerana*, however, remained in general in the pre-morphometric stage. The taxonomic units described so far are vague, changing from author to author, and no proper correlation was established between specific characteristics and geographic areas. Except for a few local analyses, diagnosis was based exclusively on small numbers of museum specimens, without taking into account intra-colony and intra-population variations. Few measurements were taken, and if so, in an inadequate and inaccurate way (e. g., "total body length" with units of 0.5 mm). *A. cerana* shows a high variability in color also within the colony, and color was the main characteristic for taxonomic diagnosis. These methods are especially unsuited for micro-taxonomy on the level of intra-specific variation. The consequence was an astonishing unsteadiness regarding the taxonomic rank of the whole systematic unit.

Buttel-Reepen (1906), who classified *cerana* (at this time "*indica*") as subspe-

Table 9.3 Morphometric data (in mm) of the "hill variety" and "plains variety" of *A. cerana* (extracted from the sources cited above; in mm)

	Tongue length	Fore wing length	Fore wing width	Cell ∅
Plains variety	4.8–5.1	7.6–7.9	2.85	4.25
Hill variety	5.2–5.5	8.5–8.8	3.10	4.80

cies of *A. mellifera*, listed six geographic varieties. The general system of grouping, maintained by most of the authors, was done according to the four regions India (*indica* F.), China (*cerana* F. and *sinensis* SM.), Japan (*japonica* Radoszkowski) and various islands of SE Asia (*johni, nigrocincta, peroni*). Skorikov (1929b) retained five units, but he raised them to the rank of species and the whole *cerana* group to a subgenus, later named "Sigmatapis" by Maa (1953). The monography of this author gives the most comprehensive review on the honeybees of SE Asia, and especially on *A. cerana*. However, the review of Maa, still based on traditional museum methods, is even more confusing than were the approaches before, since he increased the number of Skorikov's five species to 11. Some of the "new species" are based on a few worker bees only (*Sigmatapis lieftincki*, Sumatra, five individuals; *Sigmatapis vechti linda*, N Borneo, three individuals).

During the last decades numerous morphometric investigations were undertaken with local strains of *A. cerana*, esp. in India and Japan. Major result was the statement that two distinct taxonomic groups exist in India, a large, dark bee in the hills and mountains in the north of the subcontinent, and a small, yellow bee in the plains (Rahman and Singh 1950, Muttoo 1956, Kapil 1956, Deodikar et al. 1958, Narayanan et al. 1961, Ahmad 1982). These are now known as "hill variety" and "plains variety". The differences in size between the two strains are important as shown by the data in Table 9.3.

The size of honeybees in Japan and Ussuria is close to that of the hill variety (Sakai 1956, Okada et al. 1956, Lavrekhin 1967). In the Himalayas, a positive correlation was found between quantitative characters and altitude (Mattu and Verma 1980, 1983, 1984b; Verma 1984). Bee size decreases from west to east (Kashmir – Himachal – Manipur), indicating three distinct populations (Verma et al. 1984). Measuring 29 characters of populations at 22 locations all over the Indian subcontinent, Kshirsagar (1980) observed seven distinct ecotypes, five of them situated in the northern hilly region.

9.7.2 Morphometric-statistical Analysis

A more general morphometric analysis of *A. cerana* was started at the Laboratory Oberursel. The aim of this investigation was not to detect distinct local populations but to obtain an overview of the total variability of the species. 93 samples of 18 SE Asian countries, resp. regions were collected (Table 9.4). The same 40 characters were used as in *A. mellifera*. In spite of the fact that the generally known species-specific characteristics of *A. cerana* are not included in this set, a clear separation of the two species was obtained, mainly due to characteristics of wing venation (Fig. 6.9).

Table 9.4 Origin and number of *cerana* samples (total: 93)

Country or area	No. samples	Country or area	No. samples
Afghanistan	8	S Korea	5
Bali	1	Malaysia	1
Bangla Desh	1	Nepal	3
Burma	2	Pakistan	7
VR China	5	Philippines	4
India (South)	8	Sri Lanka	5
India (North)	10	Sumatra	7
Japan	8	Thailand	5
Java	9	Vietnam	4

The smallest bees found in this analysis are in the south and in the lowlands (Sri Lanka, Indonesia, Malaysia, Thailand, Philippines), the largest (and darkest) in the north and in the mountains (Afghanistan, N China, S Korea, Japan). This corresponds to the same trend found in the other *Apis* species. Differences in size are nearly as important as they are in *A. mellifera* (Table 9.5, Fig. 4.4). Another analogy to mellifera was observed: the number of hamuli taken as a character is as useless for intra-species discrimination as it is in *mellifera* (mean s.d. within geographic groups 0.539, between groups 0.395); the number of hamuli is significantly lower than in *mellifera* (15.80–19.40 vs. 18.70–25.60, Fig. 9.10). The only exception are possibly the bees of the Philippines with a lower number of hamuli (Chap. 4, p. 41).

As shown by the graphic representation of the results of a factor analysis of the morphometric data, four well-separated main groups are found in the areas examined (Fig. 9.27):

Group I: South India, Sri Lanka, Bangladesh, Burma, Malaysia, Thailand, Indonesia, Philippines. The samples of Sri Lanka are the smallest *cerana* bees measured so far, followed by the samples of the Philippines and by some of Java. A number of subclusters are established by a special analysis (Figs. 9.29, 9.30).

Group II: Afghanistan, Pakistan, North India, China, N Vietnam – a very compact group in spite of the great geographic distances and a highly disjunct area of distribution (Fig. 9.1).

Group III: Central and East Himalaya and an undefined region east of it. This group consists of seven samples from Nepal, NE India (Darjeeling and Manipur) and from the mountains in N Thailand (Chiang Mai; lowland bees from Thailand are found among group I). Moreover, the samples from N Vietnam (Hoa Binh, W of Hanoi) constitute a small cluster within group II, but quite peripherically and directed towards the Himalaya group. Surprisingly, the samples from China are not situated next to it but just at the opposite side of the cluster (Fig. 9.27). Thus these samples forming a compact cluster in the center of the *cerana* phenogram were found in mountainous sites very far from each other and it is very likely that they represent a race of bees occurring also in the mountains of N Burma and SW China. This type, being different from the mountain bee of the West (Afghanistan- Kashmir), apparently represents a bee of the tropical mountain region. In Thailand (Chiang Mai, alt. 1600 m) it is found as far south as 19° N lat.

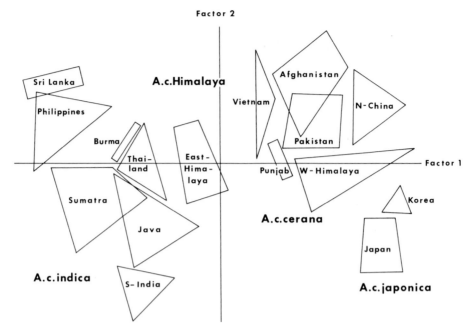

Fig. 9.27 Principal component analysis of 93 samples of *A. cerana*

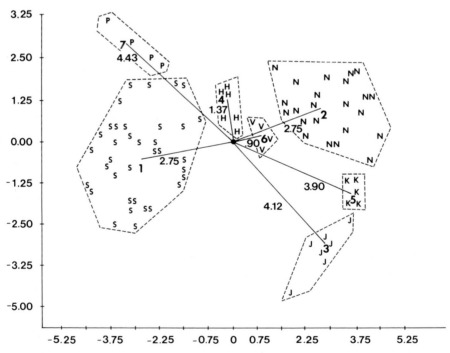

Fig. 9.28 Discriminant analysis of *A-cerana* with canonical distances 1 *(S) A.c.indica*, 2 *(N) A.c.cerana*, 3 *(J) A.c.japonica*, 4 *(H)* Himalaya, 5 *(K)* Korea, 6 *(V)* North Vietnam, 7 *(P)* Philippines

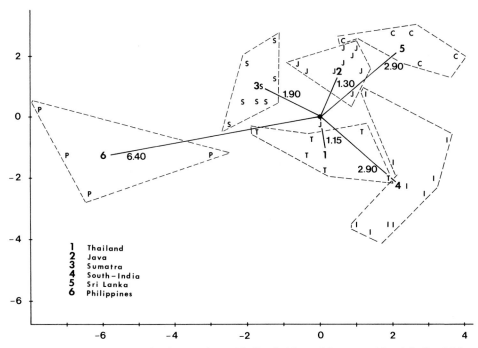

Fig. 9.29 Discriminant analysis of *A. c. indica*. *1* Thailand, *2* Java, *3* Sumatra, *4* South India, *5* Sri Lanka, *6* Philippines

Group IV: Japan, with two distinct subclusters – Honshu and Tsushima. Between this and the large cluster "N" a small compact cluster with the five samples from S Korea is found.

A discriminant analysis of the 93 *cerana* samples confirms the groups found in PCA (Fig. 9.28). Remarkable again the symmetry of the structure: "south" and "north" in exactly equal distance from the center and "Himalaya" in the middle. "N Vietnam" is now separated from the cluster N and closer to H. Separated is also the cluster "Philippines" (7) in an extreme peripheric position, exactly opposite, "Japan" (3). Again, a separate cluster "S Korea" is found between J and N.

A DA of 35 S Asian samples shows six more or less overlapping groups: Java, Sumatra and S Thailand close to the center, Sri Lanka, S India and Philippines in peripheric position (Fig. 9.29). The four samples from the Philippines (three from Luzon, one from Mindoro) are widely scattered a great distance from the center. Ellipses of confidence overlap even at the level of 75% (Fig. 9.30). Therefore, a splitting of this group into several taxonomic units seems not to be justified at the present state of analysis and more data are needed particularly for the Philippines (Table 9.5).

The drones investigated (six samples from Afghanistan, Pakistan, S India and N China) are all black with almost no yellow marks on tergites and scutellum. As in worker bees, drones from the north seem to be somewhat larger than those from the south (hind leg length – northern countries: 8.34 mm, S India: 7.56 mm).

In order to connect these results of morphometric-statistical analysis with tra-

155

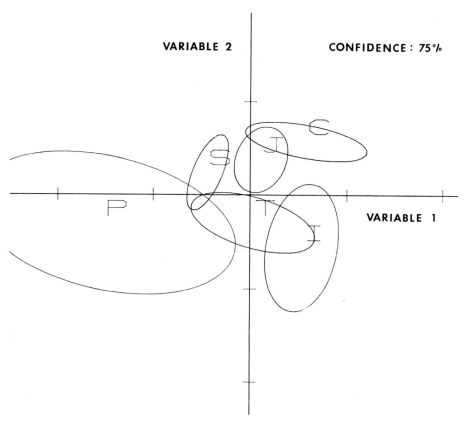

Fig. 9.30 Ellipses of confidence (75%) of the DA Fig. 9.29

Table 9.5 Limits of variation in *Apis cerana*. Measurements in mm

	Tergite 3+4	Proboscis	Fore wing	Hind leg	Color (on T3)	Cubital Ind.
max.	4.161	5.709	9.020	7.859	8.500	7.86
min.	3.182	4.410	7.472	6.435	4.500	2.63
diff.	0.980	1.299	1.548	1.324	–	5.23
%	30.8	25.6	20.7	20.6	–	198.86

ditional nomenclature, it first has to be considered that the area of *A. cerana* is not yet fully represented by this study. Essential parts such as Laos, South China, Borneo, and Celebes are not represented. For most regions only spot checks are available, not the total variation of the population. On the other hand, however, the major territories of the area are included. It is not very likely that a completely different major type remained undetected. This is certainly not the case as far as the missing island populations (Borneo, Celebes etc.) are concerned because of the recent isolation (Fig. 9.31). Therefore, at least a preliminary systematic classifica-

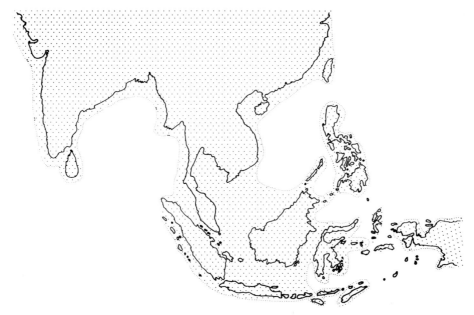

Fig. 9.31 Map of Pleistocene shore line in SE Asia. (de Lattin 1968)

tion can be attempted. Main open questions are the extension of the Himalayan bee to the east (Indochina, SW China) and the relation of the Japanese bee to the bees of the E Asian continent.

Group I, including all populations of small size (namely the "plains variety" of India and the populations of the S Asian islands) corresponds to *A. cerana indica* Fabricius 1798. Whether the bees of the Philippines should be recognized as a separate subspecies (*A. cerana philippina* Skorikov 1929) can be decided only by a more thorough knowledge of their variation.

Group II, the northern population of *A. cerana* including all the samples from N China, can be assumed to equal *A. cerana cerana* Fabricius 1793. That the populations of the western mountainous part of the *cerana* area (Aghanistan, Pakistan, N India) are not discriminated morphometrically from the bee of N China, agrees with an observation by Maa (1953).

Group III. The bees from the eastern S Asian mountains, from Nepal to Thailand and probably SW China, forming a well-separated central cluster, were recognized also by Verma et al. (1984) as a morphometrically different type.

Group IV. The complete isolation of the cluster of Japanese bees favors the proposition to retain the name *A. cerana japonica* Radoszkowski 1877 and not to include the Japanese bee in *A. cerana cerana,* as did Maa (1953). Whether the bees of Korea should be associated with this race or whether they belong to a separate, continental subspecies can at present not be decided.

The level of morphological radiation (Table 9.6) indicates that the intraspecies diversification occurred at about the same time as in *A. mellifera*: origin of major groups before or during the Pleistocene (after separation of the Japanese Islands

157

Table 9.6 Selected morphomeric data of subspecies and/or local populations of *A. cerana*. Measurements of length in mm; n number of samples

Subspecies	n	Terg. 3+4	Proboscis	Fore wing length	Hind leg	Hair	Color tergite 2	Cub. I.
cerana (Pakist.,China)	12	404.8	525.8	863.1	765.3	28.26	5.84	4.19
indica	36	346.6	478.6	778.7	682.1	17.48	7.00	3.96
Philippines	4	349.5	458.5	742.5	663.1	17.00	6.64	3.18
Himalaya	5	368.0	515.4	803.6	705.3	22.65	6.66	3.71
japonica	4	403.9	518.5	869.5	769.3	28.91	5.05	6.40
South Korea	5	410.2	528.5	847.4	767.4	28.18	7.00	5.29

from the continent), minor groups (= only slightly differentiated groups) after the Pleistocene. The islands of SE Asia have been separated from the continent for only 8000–10,000 years (Fig. 9.31).

9.7.3 Morphometric Diagnosis

Only few characteristics are needed to distinguish species, subspecies and populations:

1. Species *Apis cerana F.* – radial vein of hind wing, 4th tomentum on T6.
2. Characters for intra-species discrimination (selected by stepwise DA)

Subspecies-Discrim.		Population-Discrim. (within *A. c. indica*)	
No. of char.	Character	No. of char.	Character
7	L basitarsus	10	Terg. 4
10	Terg. 4	15	L stern. 6
11	L stern. 3	19	Vein a
13	Wax plate tr.	20	Vein b
15	L stern. 6		
19	Vein a		

9.8 Pathology

One of the most striking differences between the closely related species *cerana* and *mellifera* is their "equipment" with parasites. The species-specific mite of *A. cerana* is *Varroa jacobsoni*. It was absent in *A. mellifera* until a few decades ago – as was *Acarapis woodi* in *A. mellifera*. Both species are highly susceptible to the respective hetero-specific parasite. This implies that both specific parasitisms cannot have developed earlier than the separation of the two species.

Varroa jacobsoni Oudemans is widespread throughout the *cerana* area (Burgett and Krantz 1984). However, no severe damage is caused by this mite (Peng et al.

1987) due to a particular host-parasite relation (which is absent in general in *A. mellifera*): *Varroa* reproduces only on male brood, not on worker brood (Koeniger et al. 1980). In consequence, the parasite population remains too small to damage the bee colony. In China, Peng et al. (1987) found still another very peculiar resistance mechanism in A. cerana against Varroa mites: the worker bees remove the adult mites from their own body, from other bees, and from larvae, eliminating them from the hive. The removed mites partly show heavy damages to their bodies. Other bees are invited by a "cleaning dance" to help in the removal operation, if a bee cannot get rid of a mite within several minutes. Six or more bees were observed to work on one bee ("group cleaning behavior"). The effectiveness of this behavior was found to be very high (more than 99% removal under controlled conditions). *Mellifera* colonies, kept in the same area, showed only slight indications of this behavior, and there was no damage observed to the removed mites. It is significant that *Varroa jacobsoni* was never found in one of the sympatric species *florea* and *dorsata*; however, it switched to the allopatric *A. mellifera* wherever both species were brought together, and destroyed many colonies of this species. *Tropilaelaps clareae*, considered as a specific parasite of *A. dorsata*, was found also infesting *A. cerana* colonies (Delfinado-Baker 1982; Ahmad 1984; Verma 1985); however, no damage was reported.

Acarapis woodi is found in northern India since 1956, when *A. mellifera* was imported on a larger scale, "spreading like wild fire" in India and Pakistan and causing severe problems since then (Atwal and O.P. Sharma 1971; Dinabandhoo and Dogra 1980; Kshirsagar 1982; Verma 1983). In Pakistan, 8000 *cerana* colonies died in the years 1982/83 in consequence of infestation by this mite (Ahmad 1984). Incidence of *Nosema apis* Zander was reported from north India (L.R. Verma 1984). During experiments in the flight room of the Apicultural Institute in Oberursel, *A. cerana* turned out to be nearly *Nosema*-proof. In contrast to *A. mellifera*, no permanent treatment with a drug was necessary. Every few weeks a sample of bees was examined and only by the end of the winter a few solitary spores were found. Sac brood, a viral disease, has been observed in *A. cerana* only since the importation of *A. mellifera*. It occured first in Thailand, and is now causing severe losses of colonies in northern India (Verma 1985). Another imported dangerous pest of *A. cerana* is foulbrood *(Bacillus larvae)*, (Kevan et al. 1984).

Wax moths (*Galeria mellonella* and *Achroia grisella*) infest the combs of all *Apis* species. In *A. cerana* they attack especially weak colonies and cause them to abscond (Ahmad et al. 1985; Fang Yue-Zhen 1984). The favorable conditions for wax moth's growth in *cerana* nests are caused by the inherited trait of dismantling old combs (p. 132) combined with poor cleaning behavior. The wax litter on the bottom of the nest provides an ideal substratum for undisturbed rearing of wax moths. This noxious situation may be regarded as one of the inconsistencies of evolutionary trends. Or could it be a relic of a free-nesting ancestor, who just dropped wax particles on the ground, without deadly consequences for the colony?

9.9 Apis cerana and Man

Beekeeping tradition with *A. cerana* in eastern Asia is probably as old (2500-2700 B.C. in India; Joshi et al. 1980; see also Muttoo 1956; Sakagami 1958; Ma De-Feng and Huang Wen-Cheng 1981) as apiculture with *A. mellifera* in Egypt, the Near East, and Europe. Also the material used to make containers to hive the bee colonies were essentially the same: pots made of clay of different size and shape, wooden or wicker boxes, log hives, niches in walls (Schneider P. and Djalal 1970; Nogge 1974; Sakagami 1972).

Beginning this century (somewhat earlier in Japan), movable frame hives were introduced, in some regions as India with reduced size (Fig. 9.32), adapted to the size of *cerana* colonies: "Newton Hive" widespread in southern India and "Jeoli-kote Villager Hive" in the north (Muttoo 1956). Hives used in South India have eight small frames (13 × 21 cm) and two shallow supers Reddy 1980).

However, even when modern standardized methods (e.g., Langstroth hive) are applied, honey yields are substantialy inferior with *A. cerana* as they are in the same environment with *A. mellifera*. It is not surprising that *A. mellifera* was and still is imported in practically every country of the *cerana* area. In some countries

Fig. 9.32 *Cerana* with small movable frames. (Photo Lindauer)

160

as in Japan, China, Northern India, and Taiwan the initial difficulties were overcome and a permanent *mellifera* apiculture has been firmly established, gradually completely ousting *A. cerana*. In the PR China, out of six million hive colonies, only about one million was *A. cerana* in 1984, predominantly in the S and SW of the country (Fang Yue-Zhen 1984). This proportion is similar in South Korea (Song-Kun Lee, pers. commun. 1985). In Japan, *A. cerana* is found only in remote mountainous areas, almost like a relic which could soon become an "endangered species".

In other (especially tropical) countries, however, difficulties with imported *mellifera* colonies continue to exist, mainly due to adopted parasites (*Varroa, Tropilaelaps),* predators such as wasps and birds and mating competition by *A. cerana* (FAO Consultation 1984). Tragically, the impact of these problems is mutual: *A. cerana* is suffering severely from acquired parasites and diseases of its newly imported sister species: acariosis, sacbrood, foulbrood. Apiculture has become more complicated in these regions, and no final solution is in sight.

Part II   The Western Honeybee Apis mellifera L.

Classification and Natural History

CHAPTER 10

Apis mellifera Linnaeus 1758:576
General Introduction to a Polymorphic Species
with an Unusual Range of Adaptation

In this species all choices seem to be tested which are possible in the multi-comb variation of the basic *Apis* type: many kinds of climates ranging from cold temperate to tropical, from permanent humidity to semi-deserts; several brood patterns, genetically adapted to different floral seasons; different rates of reproduction, ranging from long-living colonies with low reproductivity to short-living, frequently reproducing colonies.

10.1 Types from the Tropics and the Temperate Zone

The most obvious difference exists between *mellifera* and *cerana* races of the tropics and of the temperate zone (Chap. 1). Accepting the general assumption that all *Apis* species are of tropical origin, both cavity-nesting species (the only ones among all temperate-zone insects; Seeley and Visscher 1985) show the potential for change from a life strategy best suited for the tropics to characteristics adapted to the temperate zone: abolishing the tendency to abscond; reducing the number of swarms and shifting the swarming period to the first half of the season; building up colonies with large populations and ample provisions. In *A. cerana*, colonies of northern origin have a low tendency to abscond and also to swarm (pers. observ.), but no scientific data exist to assess this on a larger scale. *A. mellifera*, on the other hand, is fairly well known through several subspecies, as will be shown in the following chapters. The study of honeybee races called attention to a number of features essential for their existence in specific environments.

It is obvious that a population lives in balance with the resources of the environment. If an unmanipulated population shows a high swarming tendency, this indicates either the loss of colonies is high (at least periodically) or only some swarms succeed in establishing new colonies. A low swarming tendency should indicate the opposite. Seeley (1985) extensively discusses the theory of colony fission primarily considering the aspects of kin selection. Reviewing the natural conditions in different environments, the problem appears to be quite complex.

Ample food supply enhances rapid colony growth, then follows overcrowding with subsequent swarming and afterswarming. In the temperate zone selection limits the swarming period: too many and too late swarms cause death for both mother colony and the swarms. Therefore, the swarming tendency is in part genetically fixed and even several hundred years in the tropics of S America were not able to change the low swarming tendency of Iberian bees (Winston et al. 1983). The high rate of swarming of Africanized bees in the same region indicates, in addition to a "tropical" genome, an unsaturated environment. It would be of interest to study the swarming frequency of an established Africanized population when food and nesting site competition is limiting colony growth. As shown in breeding programs, a wide genetic variability exists within natural populations which are open to selection in both directions (Ruttner 1983). The highest rank of a genetically determined swarming impulse is a prime swarm in spite of empty space and secondary swarms of an afterswarm (in German "Jungfernschwarm"; Berlepsch 1873).

In *A. mellifera*, the limiting factor of feral population size is mostly lack of suitable nesting sites rather than lack of food. This is the price for the need of a protective cavity, in contrast to the open-nesting species. In the African savanna empty boxes are rapidly occupied by swarms during the appropriate season. Many of them have to be content with substandard sites, too small for a full sized-colony: mounds of termites, crevices in rocks, small cavities in trees. Consequently, rapid overcrowding and new swarms occur. Nevertheless, a high reproduction rate whenever permitted by the conditions is needed to make up for unpredictable losses due to years of drought, bush fires, etc. Also very small swarms are able to survive and to develop to full strength.

The habitat of honeybees in the temperate zones are cavities in deciduous trees which are likely to be large enough to allow for the growth of large colonies (Seeley and Morse 1976). Swarming is a special risk in regions without a late nectar flow. Seeley (1978) and Seeley and Visscher (1985) found a distinct difference in the survival rate between early and late swarms (which had only little chance to come through the winter). The first author found a reproductivity of 0.9 per feral colony in New York state, a figure which is likely to vary among years and regions. Due to highly variable weather conditions in the temperate zone, winters with high colony losses and summers with increased swarming tendency alternate irregularly. It is the mean over a number of years which gives the correct population balance.

10.2 Overwintering

Only a few of the known races of honeybees were able to evolve characters which permit survival during long, hard winters with a minimum of risk. These are *mellifera* and *carnica* in northern and alpine strains and probably *anatoliaca* (Br. Adam 1983). Notwithstanding the seemingly harsh conditions in higher latitudes, the survival chances of well-adapted races are evidently not diminished compared with southern regions.

A number of characters are essential for wintering of a bee colony:

1. Individual Size. The two races in the most northern *Apis* area are the largest of the species (except for A. m. major, a peculiar local population of North Africa). "Endothermy is a question of size" (Bartholomew 1982).

2. Efficient Thermoregulation. The short season in higher latitudes necessitates the start of brood rearing while very low ambient temperatures prevail. Differences of more than 60°C between brood nest and ambient temperature were measured (Seeley 1985). Experimentally, Southwick demonstrated the ability of a bee colony (1.6 kg of bees) to maintain, without any difficulty, homeothermy for at least 12 h

Fig. 10.1 Bees on the surface of a swarm, *a* (*top*) at 25°C and *b* (*bottom*) at 3°C (Heinrich 1985)

at a temperature of −80° C. The core of the cluster kept steadily at 34° C, the peripheric layer of bees measured 9° C.

At least four factors are involved in this cooperative accomplishment. a) A tight, contractible cluster which reaches its minimum size at 0-2°C (Southwick 1985a). During low ambient temperatures head and thorax of surface bees are bent inwards, the "cool" abdomen is exposed (Fig. 10.1b, Heinrich 1985). The cluster is much more efficient in heat production than the sum of the individual metabolism rate. b) Low conductivity of the cluster (comparable to fur of mammals and feather cover of birds). The plumose hairs of the clustered bees interlace and trap a large volume of air (Southwick 1985b). The insulating properties of tightly clustering bees are reinforced by the wax comb with its empty cells (Fig. 10.2); Beekeepers know that an "overfed" colony will not survive if the comb cells at the site of the cluster are filled with honey. c) Metabolic heat production by individual bees (Southwick 1987). d) Mass shifting from the outside into the warm center when bees cool to 9° C.

3. "Winter Bees" are physiologically different from summer bees: being exempted from brood care, they have accumulated protein and fat in the food glands and the subdermal layer of the abdomen through rich nutrition (including biopterin which is missing in the brood food of summer bees; Hanser and Rembold 1960). The indispensible capacity of the rectum to store large quantities of feces seems to be improved by an increased catalase production by the rectal glands in autumn.

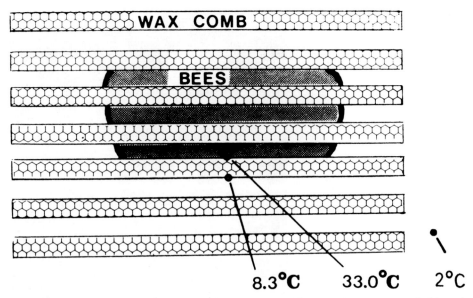

Fig. 10.2 Insulating effect of an empty comb: temperature difference of 25°C between inside and outside. (Southwick 1985 b)

This was found to be double in bees of the northern USSR (Moscow, Novosibirsk) than in bees of Tashkent and Maikop; bees of Tashkent transferred to the north did not increase the catalase production in autumn as did the local bees (Sherebkin 1976).

4. Low (and Early) Swarming to secure a sufficient winter population of 12,000–20,000 bees with ample stores. In an experiment conducted in New York State over a period of 3 years it was found that in a medium year early swarms survived the next winter while late swarms perished (in very bad years all swarms died; Seeley and Visscher 1985). From data on feral colonies in the northeastern US it was calculated that a swarming rate of 0.9 is sufficient to maintain the population (Seeley 1978).

5. Long-Lasting Period without brood production during winter. This keeps food consumption and, consequently, the strain on the rectum at a low level. Winter food consumption (October to March) of *carnica* colonies in Germany was determined over a period of many years to average 5.2 kg (Zander-Weiss 1964). The total amount of winter stores needed from August to the first honey flow in spring in Central Europe is traditionally given as 10–12 kg. This is only half the quantity *ligustica* hybrids need in the northeastern US. They produce a high amount of brood in autumn and have only a short diapause until midwinter (Seeley 1985). *Carnica* colonies overwinter usually at a smaller size, but they increase more rapidly in spring than Italians (Fig. 10.4; Br. Adam 1983; Ruttner 1983). Seeley and Visscher reported negative effects of experimentally inhibiting winter brood, but these conditions may substantially differ from those in an undisturbed, well-adapted colony. According to apicultural experience, not fully winter-adapted

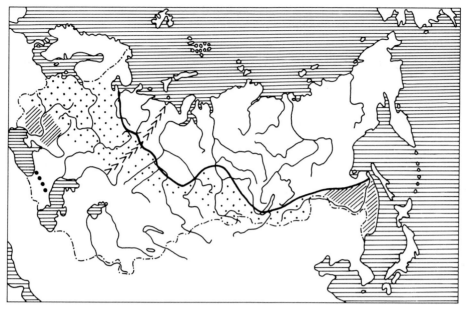

Fig. 10.3 Northern limit of honeybees in the USSR (*dotted line*). Area of *A.m.mellifera: dotted.* Ukrainan bee: *hatched.* (Alpatov 1976)

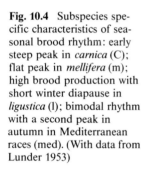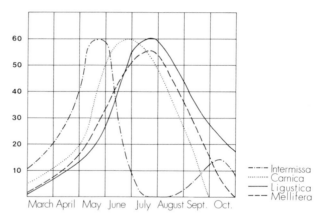

Fig. 10.4 Subspecies specific characteristics of seasonal brood rhythm: early steep peak in *carnica* (C); flat peak in *mellifera* (m); high brood production with short winter diapause in *ligustica* (l); bimodal rhythm with a second peak in autumn in Mediterranean races (med). (With data from Lunder 1953)

races of the warm temperate zone, like *A.m.ligustica* and *cypria,* can also safely survive long winters, provided certain precautions are taken (selection, artificially strong colonies).

6. A Certain Level of Resistance against nosema disease. Most likely there is no complete resistance in *A.mellifera* as in *A.cerana,* but differences in susceptibility were demonstrated (Steche and Böttcher 1984). According to Bilash et al. (1976), Ruttner 1983 and others, wintering of *A.m. caucasica* is poor in cold climates and heavy losses may occur because of this disease.

169

Successful "wintering" not only means tolerance of cold winter months, it also includes coping with changing, frequently harsh conditions from March to the middle of May. The death rate is greatest at the time when old bees with a sometimes spreading *Nosema* infection have to cope with the stress of high brood rearing activity. Strains which genetically respond correctly to climatic conditions are the best adapted. Differences are found not only between but also within races (Louveaux 1969).

The limits of winter adaptability are reached in NE Europe. According to Alpatov (1976) bee colonies can survive where rivers are frozen for less than 6 months (Fig. 10.3). It is, however, not quite clear whether this also includes colonies in feral conditions. Honeybees in Scandinavia have been autochthonous in South Sweden but not in Norway. At present economic apiculture is practised in Norway up to 67° N.lat. (Villumstad pers. commun.).

The reward for this dangerous wintering effort is a summer paradise for flower-visiting insects. The short season in the north is compensated for by longer daily flight activity and a higher nectar production induced by the increased daylight length (Shuel 1966). M. Lindauer transported a nucleus of *carnica* bees (only very modestly provisioned) for experimental purposes from Frankfurt to Abisco on the polar circle. In only 3 long days this nucleus was completely full of honey. In the northernmost district of Sweden, Norbotten, bees are kept and hibernating on the polar circle. High honey yields are generally achieved there (21–37 kg) and in 7 out of 30 years (1940–1969) this district recorded the highest production in all of Sweden (Roswall 1974). Honey production in eastern Siberia (Transbaikalia), in an extreme continental climate (winter minima to $-50°$ C), is especially high (70–80 kg; Kotzin 1931). In the Amur district average honey yields of well above 100 kg were recorded, with daily weight gains up to 30 kg (Danilenko 1975). The main supplier of bee wax and honey to Central Europe during the 14th century was Russia (especially Bashkiria), according to the market reports of Brügge, the center of the trade (Büll 1961). From these data high productivity of the traditional northern forest apiculture can be deduced.

In many places in the cold temperate zone it was experienced that total honey amount needed for winter could be stored within a few days. But not a single one of a number of races or populations imported for experimental purposes during more than 30 years at the Institutes of Lunz am See (Austrian Alps) and Oberursel showed overwintering qualities equal to the local strains of *carnica* and *mellifera*. Problems with wintering colonies in the northern regions of N America most likely originate from a not fully adapted bee (*ligustica* and hybrids), as indicated also by published data on their biology (Seeley 1985).

10.3 The Brood Cycle

The brood cycle is symptomatic for the timing of the seasonal activity. The invention of the movable comb opened the possibility of quantifying the amount of brood. It was shown that the different types of brood rhythms (Fig. 10.4) are correlated with the floral cycle of the region, but that they are also permanent characteristics of the local bee strains (see p. 239; Louveaux 1969). It is of interest that different types of brood rhythm may occur within one and the same race (e.g., *A.m.mellifera* in France) or that the same rhythm is found in several races of a certain region (e.g., the Mediterranean, Fig. 10.4). Usually, if an imported strain shows poor performance it is because of an inherited brood rhythm not fitting in with the floral rhythm of the region.

10.4 Reproduction

The mating of the queen (of the mother colony after swarming and of the virgins in the afterswarm or after supersedure) is one of the main hazards in the colony's life. In apiculture 20% loss during the mating flight is considered to be normal, and this might not greatly differ from feral conditions. Units without queen and brood occasionally might be saved by parthenogenetically produced females (Butler 1954), but this occurs very rarely. The price of considerable losses is apparently not too high to secure outbreeding. Nevertheless, two different ways are observed to reduce these losses:

1. Habitual Production of females by worker bees (thelytoky) in the Cape bee (p. 217). Isolated attempts of parthenogenetic reproduction are frequent in the animal kingdom, but they evidently remain without lasting success. This seems to be the case also in honeybees, although in Cape colonies queenrightness is restored in most cases (Ruttner 1975).

2. Polygyny during the mating period: several virgins coexist peacefully in the swarm or the mother colony, until one virgin is mated and monogyny is restored. This condition is found with a number of Mediterranean races *(syriaca, lamarckii, intermissa, sicula)*. Since this seems to be a very simple and efficient solution of the problem, it is hard to understand why this system was not generally adopted by *A.mellifera*. Perhaps it is a relatively new acquisition with regional distribution only, or it has some disadvantages not yet understood.

10.5 List of Geographic Races

Apis mellifera is a highly polytypic species and many of the general aspects of its variability are discussed in Chapter 4. Most varieties easily discriminated by their general aspect have been recognized as subspecies for a long time. A clear classification of the whole species, however, was achieved only by a morphometric survey of all accessible regions of the *mellifera* area, using a set of carefully selected characters. In consequence, several taxa were suppressed. Twenty four taxonomic units can be discriminated morphometrically at the present state of knowledge. They correspond to the criteria of a geographic race as discussed in Chapter 6. One of them, *A.m.armeniaca*, is represented only by six samples in our analysis and very close to *A.m.anatoliaca* in a special DA (Fig. 11.1) but well separated in the over-all analysis of workers and drones. A possible 25th race, described as *A.m.major* (Ruttner 1975), is not yet fully understood in its evidently restricted

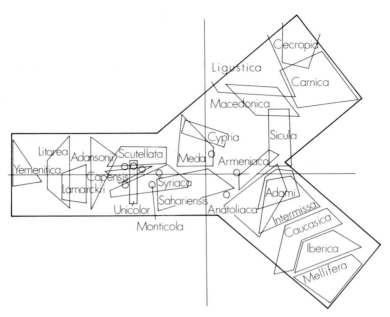

Fig. 10.5 Graphical presentation of a principal component analysis of all 24 races. n = 220 worker samples. Horizontal: factor 1, vertical: factor 2. Clusters schematized. Circles: samples of *monticola* not forming a cluster

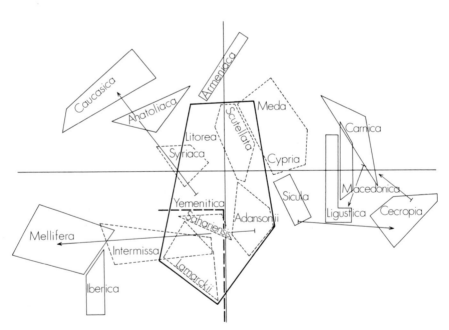

Fig. 10.6 PCA same as Fig. 10.5. Horizontal: factor 2; vertical: factor 3. The fourth branch (*O*-branch) is visible with *caucasica* at the end. Heavily lined: *A*-branch with races of tropical Africa. Arrows: "Rassenkreise" corresponding to *M, C,* and *O* branches

area in the middle of the common N African bee, *A.m.intermissa,* although it is very distinct morphometrically. Another small local population described as a geographic race is *A.m.taurica* (Alpatov 1938b) from the Crimean peninsula. The first description being rather preliminary, this unit canot be given equal status with the other races until further data are provided.

At this time 24 distinct taxonomic groups (races) can be discriminated by the morphometric methods described earlier. They are classified into three groups for convenience and not necessarily with the intention of indicating a phylogenetic relationship. A further splitting would result in artificial units, the subspecies being arranged in a net-like pattern (Fig.10.7) which is hard to transform into a linear system.

I. Near East

    1. *Apis mellifera anatoliaca* Maa (1953)
    2. *A.m.*   *adami* Ruttner (1975)
    3. *A.m.*   *cypria* Pollmann (1879)
    4. *A.m.*   *syriaca* Buttel-Reepen (1906)
    5. *A.m.*   *meda* Skorikov (1929a)
    6. *A.m.*   *caucasica* Gorbachev (1916)
    7. *A.m.*   *armeniaca* Skorikov (1929)

II. Tropical Africa

    8. *Apis mellifera lamarckii* Cockerell (1906)
    9. *A.m.*   *yemenitica* Ruttner (1975)
  10. *A.m.*   *litorea* Smith (1961)
  11. *A.m.*   *scutellata* Lepeletier (1836)
  12. *A.m.*   *adansonii* Latreille (1804)
  13. *A.m.*   *monticola* Smith (1961)
  14. *A.m.*   *capensis* Escholtz (1821)
  15. *A.m.*   *unicolor* Latreille (1804)

III. Mediterranean

  1. West Mediterranean
    a) North Africa
      16. *Apis mellifera sahariensis* Baldensperger (1924)
      17. *A.m.*   *intermissa* Buttel-Reepen (1906)

    b) West Mediterranean and North Europe
      18. *A.m.*   *iberica* Goetze (1964)
      19. *A.m.*   *mellifera* Linnaeus (1758)

  2. Central Mediterranean and Southeast Europe
      20. *A.m.*   *sicula* Montagano (1911)
      21. *A.m.*   *ligustica* Spinola (1806)
      22. *A.m.*   *cecropia* Kiesenwetter (1860)
      23. *A.m.*   *macedonica* Ruttner (1987)
      24. *A.m.*   *carnica* Pollmann (1879)

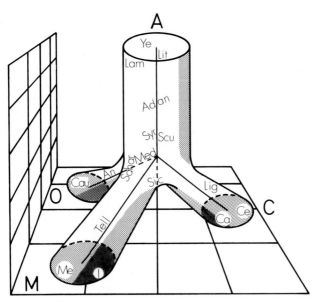

Fig. 10.7 Three-dimensional phenogram of *A.mellifera* with *M, C, A,* and *O* branches. *Adan adansonii, An anatoliaca, Ca carnica, Cau caucasica, Ce cecropia, I iberica, Lam lamarckii, Lig ligustica, Lit litorea, Me mellifera, Med meda, Sahar sahariensis, Scu scutellata, Sic sicula, Syr syriaca, Tell intermissa, Ye yemenitica*

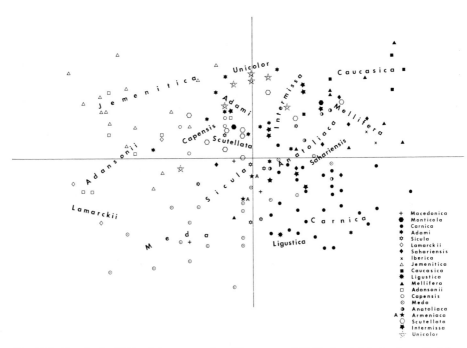

Fig. 10.8 Graph of a PCA of drone samples; 20 races, n = 162. Horizontal factor 1, vertical factor 2

10.6 The Morphometric Structure of the Species

If samples of all 24 races are simultaneously processed in a PCA, the clusters arising by factor 1 and 2 are arranged in the shape of a horizontal "Y" (Fig. 10.5). The stem of the letter is occupied by the races of tropical Africa with *yemenitica* and *litorea* at the end. Of course, clusters tend to overlap highly in this overall analysis. The *lamarckii* cluster (Egypt) is situated clearly among the bees of tropical Africa. At the borderline the clusters *scutellata* and *unicolor* are next to *syriaca* and *sahariensis*. *Monticola* is overlapping with Mediterrenean races. One of the branches of the "Y" is formed by the Balcan races with *cecropia* and *carnica* at the outermost (C-branch), the other by the two West European races *iberica* and *mellifera* (M-branch). Surprisingly, the *caucasica* samples, very similar in their general aspect to *carnica*, are found in the M-branch very close to *iberica* and *mellifera*. However, in a PCA with factors 2 and 3 (Fig. 10.6) and in a tridimensional aspect (Fig. 10.7) *caucasica* appears at the end of a separate fourth branch (0-branch). *Armeniaca* is clearly separated from *anatoliaca* in this analysis (see p. 198). In the area of ramification the medium-sized oriental and Mediterranean races are found, with much overlapping in this analysis.

The mophometric analysis includes 162 samples of drones, using 24 characters (p. 74). Included are all races, except *A. m. cecropia, cypria, litorea* and *syriaca*. The arrangement of the drone samples in a PCA is in principle similar to the arrangement of worker samples, but instead of three, four poles in the plane of coordinates 1 and 2 are formed: *yemenitica - caucasica* and *mellifera - carnica - meda* (Fig. 10.6). Races of tropical Africa are found left of an oblique line consisting of the clusters *meda - sicula - intermissa*. In several cases the morphometric arrangement is not consistent with possible geographic or phylogenetic relations: *adami* drones (from Crete) are situated between the clusters *unicolor* and *scutellata*, *sahariensis* between *carnica* and *mellifera*. The two *armeniaca* samples are well separated from the *anatoliaca* cluster. *Caucasica* and *mellifera* are almost united in one cluster.

In 151 samples characters of workers and drones were tentatively analyzed together (47 characters). The arrangement of the samples in a PCA is similar to Fig. 10.5 (workers alone), with a somewhat better discrimination.

CHAPTER 11

The Races of the Near East
(Irano-Ponto-Mediterranean Area)

11.1 General

The taxonomic study of honeybees from the Near East and the Caucasian region has been a confusing puzzle in the past, rather than solid scientific work. Some varieties were named but not properly described more than 100 years ago, but even that was not performed according to the rules of nomenclature: the still frequently used name "remipes" for a bee of the Caucasian region was first published in 1862 by Gerstäcker and reused in another sense by Skorikov (1929a), both referring to a letter, or rather to an unpublished manuscript by Pallas.

In 1889, Pollmann described Caucasian bees imported to northern Russia and Germany as "*Apis mellifica caucasia*. Evidently bees of the same origin were named and biometrically characterized as *A. m. caucasica* by Gorbachev in 1916. To complete the confusion, several local races were established in the Caucasus mountains by Skorikov (1929) based only on differences in color (a highly variable character in this region) and in tongue length, namely Grusinian, Abchasian, Mingrelian, Imeretican bees.

The early taxonomic attempts followed the apicultural reputation after repeated importations of Caucasians to Europe and America. The same happened with other races of the Near East, e.g., *A. m. cypria*. The bees of Anatolia were given publicity because of the travels of Brother Adam through Turkey in 1962 and 1972. Other races of the region, however, remained almost unknown and unnamed (such as the bees of Iran).

It is hardly useful to describe in detail the earlier attempts at classification of these bees. Single exceptional studies are Bodenheimer's "Bees and beekeeping in Turkey" which will be mentioned later. In the meantime, however, the situation was changed by coincidental favorable circumstances: Brother Adam made available his collection of bee samples from Anatolia, Lebanon, Israel, and Jordan for biometrical analysis; Dr. D. Pourasghar, former student at the Institute in Oberursel, collected a great number of samples in all parts of Iran; the Soviet Ministry of Agriculture together with G. Bilash (Apicultural Institute Rybnoe) provided an ample collection of bees from various locations in the Caucasian and Armenian region. Including samples of other origin, a coherent taxonomic analysis of this particular *mellifera* region can finally be presented.

11.1.1 Climate

Only one common climatic factor exists for the whole of SW Asia: warm (to extremely hot) summer throughout the whole area; but in all other factors great contrasts are found: a mild, humid winter everywhere on the coast, but long-lasting frost and snow in the central highlands of Anatolia. The area of Erzerum is covered by snow for 4 months, and the mean January temperature is as low as −1.8° C. In summer, the same regions reach maximum temperatures of 40°C and higher (Brice 1978). All types of vegetation are found: subtropical flora with palm and citrus trees, mountain forests, dry steppes in the upland with a short flowering season, true deserts, and alpine grasslands. Aboriginal bees are found everywhere, except in the rockless steppes and the true deserts. Therefore, as far as the ecological conditions are concerned, a common denominator can be found for this group of related races only within restricted limits.

11.2 Geographic Variability

11.2.1 General

Surprisingly, morphometrically the same bee can be found in most contrasting climates, e.g., in the subtropical forests of the Caspian coast and the high mountains of central Iran, or on the coast of the Black Sea and the Caucasian alpine region. On the other hand, sometimes a borderline is found where one race is replaced by

Table 11.1 Morphometric characteristics of honeybee races of the Near East Mean (measurements in mm) and standard deviation (sd)

Sub-species	n	Terg 3+4	Proboscis	F wing length	Hind leg	Basit-I	Hair length	Toment-I	Color terg 2	Cub-I
anatoliaca	40	4.464	6.462	9.188	8.095	57.61	0.290	2.13	5.36	2.24
sd		0.091	0.169	0.134	0.135	0.99	0.032	0.26	1.43	0.18
adami	24	4.526	6.460	9.085	8.195	56.38	0.301	3.06	5.62	1.89
sd		0.089	0.105	0.040	0.134	0.81	0.031	1.02	2.37	0.18
cypria	8	4.237	6.390	8.865	7.875	57.82	0.266	2.27	8.63	2.72
sd		0.163	0.135	0.149	0.178	2.14	0.028	1.06	0.54	0.36
syriaca	11	4.112	6.191	8.482	7.828	56.32	0.226	2.39	8.79	2.28
sd		0.150	0.226	0.228	0.196	2.04	0.040	1.08	0.42	0.37
meda	93	4.356	6.335	8.967	7.821	56.36	0.285	2.31	8.27	2.56
sd		0.149	0.211	0.231	0.225	2.75	0.041	0.76	1.34	0.72
caucasica	27	4.547	7.046	9.319	8.296	57.68	0.335	2.79	3.80	2.16
sd		0.118	0.189	0.183	0.180	2.10	0.031	0.40	0.78	0.31
armeniaca	6	4.498	6.646	9.068	8.060	57.17	0.326	2.70	8.78	2.61
sd		0.109	0.114	0.144	0.166	2.00	0.031	0.34	0.52	0.42
carnica O	21	4.515	6.396	9.403	8.102	55.59	0.288	2.04	1.89	2.59
sd		0.123	0.154	0.150	0.166	1.80	0.063	0.45	1.07	0.42

another without any substantial change in climate and vegetation, e.g., in eastern Anatolia. Isomorphology does not necessarily mean the same ecological adaptation and behavior. There is abundant evidence regarding ecological diversity within the same race (Louveaux 1969) and there are indications that also in Anatolia the bees on the coast and in the central highlands, though belonging to the same taxonomic unit, are different in their physiology and behavior (Br. Adam 1983).

Rules of zoogeographic variation (tall, dark and long-haired varieties in the north, and small, yellow and short-haired types in the south, p. 52) can be verified only generally and with exceptions: *syriaca* and *cypria* are distinctly smaller than the northern races and very yellow, but *adami* of the same geographic latitude is as large as *caucasica* and the northern *armeniaca* as yellow as *syriaca* (Table 11.1). In the Caucasus, samples were compared from different altitudes, from 100 m to 1800 m (Table 11.3).

SW Asia is a zone of high morphological diversification and evolution within the species *Apis mellifera*. In this isolated part of its area, a group of clearly distinct races evolved, adapted to extreme varieties of external conditions. Genetic center of this group is Asia Minor, as revealed by statistical analysis (Fig. 11.2), with extreme climatic contrasts, both in its history as in its present status.

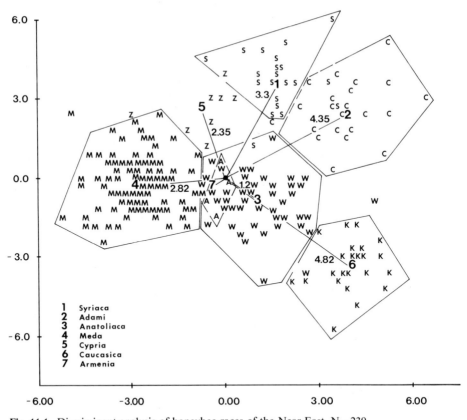

1 Syriaca
2 Adami
3 Anatoliaca
4 Meda
5 Cypria
6 Caucasica
7 Armenia

Fig. 11.1 Discriminant analysis of honeybee races of the Near East. N=239

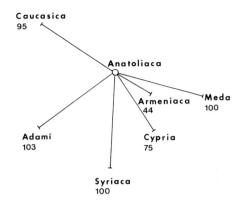

Fig. 11.2 Relative distances between centroids of races, *A. m. anatoliaca* as the center

11.2.2 The Major Taxonomic Units of the Region

The application of different methods of multivariate statistical analysis (e. g., PCA) resulted in the discrimination of six well defined races. A seventh possible type, located in Armenia, forms a compact cluster, but this overlaps with the cluster of *A. m. anatoliaca*.

The Near East, including the Caucasus, represents phenetically an almost totally isolated part of the whole *mellifera* area. Morphometrically, only two races – the western population of *A. m. anatoliaca* and *A. m. syriaca* – show (relatively distant) relations to adjacent groups. The other races, especially *caucasica, meda*, and *adami* are positioned peripherally in the statistic-morphometric structure of the species. By discriminant analysis the same six well separated clusters are obtained (Fig. 11.1). The samples from Armenia (Kartshevan, Megrinski region) are found between the clusters "*meda*" and "*anatoliaca*", partly overlapping with the latter and looking like a local population of this race. Although only five worker and two drone samples from a single location are at disposition for analysis, we preliminarily will treat this bee as a separate taxonomic unit; the reason being that the Armenian bee was frequently mentioned as a distinct race (Alpatov 1929; Markosjan et al 1976; Avetisyan 1978) and that in the total PCA of all 24 *mellifera* subspecies, workers as well as drones, the Armenian samples are clearly separated from the *anatoliaca* cluster (Figs. 10.6, 10.8). Only a wider-ranging knowledge of the variation and distribution of this type will reveal whether this classification is correct or not.

Measuring the statistical distances between the centroids of the clusters and putting the distance *anatoliaca* - *meda* = 100 (Fig. 11.2), it surprisingly becomes evident that three other races (*syriaca, adami, caucasica*) are located almost exactly in the same range. The centers of both *cypria* and *armeniaca*, however, are somewhat closer to the common center, *A. m. anatoliaca*. Since the western population of *anatoliaca* (from the region of Istanbul - Bursa - Izmir) has morphometric similarities with bees of SE Europe (*A. m. macedonica*, p. 249), the central Mediterranean (*A. m. sicula*) and north Africa, this race can be regarded as the eastern genetic center of *A. mellifera*.

The centroids of the races *caucasica-armeniaca* and *caucasica-meda*, although close neighbors geographically, are far more distant from each other (138 and

198 units, respectively) than any single one of these races from *anatoliaca*. (Fig. 11.2)

As far as size is concerned, *A.m. syriaca* is the smallest bee of the Near East and *A.m. caucasica* the largest (Table 11.1). All races with exception of *caucasica* have predominantly yellow banded bees. *Syriaca, cypria, meda,* and *armeniaca* have uniformly the highest color classifications (nos. 7–9) for tergite 2 and 3, that is bright yellow bands with narrow (or even almost absent) dark stripes.

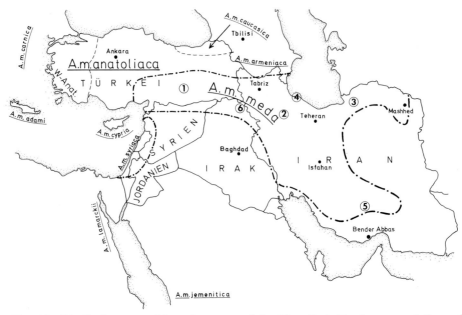

Fig. 11.3 Distribution map of honeybee races of the Near East. Numbers = populations of *A.m. meda* (*1* SE Anatolia; *2* central Iran; *3* NE Iran; *4* Caspian coast; *5* S Iran; *6* Iraq)

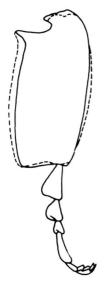

Fig. 11.4 Basitarsus of worker hind leg of *A.m. caucasica* (broken line) and *A.m. mellifera* (solid line)

For all bees of the Caucasian region (including Armenia) Skorikov (1929a) created a separate species, *Apis remipes* Skor. The name "remipes" ("Oar footed"), already used by Pallas and Gerstäcker, was supposed to express the fact that these bees have broad basitarsi (and tibiae) of the hind leg. This character is measured by the basitarsal index, BI (width:length × 100), Fig. 11.4. It is true that *caucasica* has broad basitarsi (BI = 57.68) compared with European races (*ligustica* 55,34, *carnica* 55.59, *mellifera* 55.41). If all races with a BI higher than 57.00 are classified as "remipes", not only *armeniaca* (BI 57.17) has to be included in this species, but also *anatoliaca* (BI = 57.61) and *cypria* (BI = 57.82). Moreover, we also find *yemenitica* (57.14), *unicolor* from Madagaskar *(57.19)* and several samples of *adansonii* (overall mean 56.97) in this group. Therefore, only limited significance can be attributed to this character alone. This example again demonstrates that intra-specific taxonomic classification usually cannot be restricted to one or two characters only.

The bees of the Near East were highly isolated from the rest of the species before interference by man. North of this area, no feral colonies can survive because of the lack of nesting sites and food in the dry steppes of south USSR; to the west (Ukraine) bees were introduced only 500 years ago, to the east (Wolga and Kazachstan) only during the last century (Skorikov 1929). In recent times no bees exist along the eastern border of Iran (Figs. 9.1 and 9.4) and in the Syrian-Arab desert or in the Sinai. All the other borders are sea coasts, except for a short line of contact with European bees in Thrace.

11.2.3 Statistical Characteristics of Oriental Honeybee Races

Discriminant variables extracted by stepwise discriminant analysis

No. of char.	Character	No. of char.	Character
2	Tomentum 4 (width)	19	vein A, length
6	Tibia (length)	20	vein B, length
7	Tarsus (length)	24	ven. angle E9
9	Tergite 3 (longit.)	25	ven. angle G18
10	Tergite 4 (longit.)	26	ven. angle I10
14	Distance of wax mirror	28	ven. angle K19
18	Fore wing (width)	29	ven. angle L13
33	Color tergite 3	30	ven. angle N23
35	Color scutellum	31	ven. angle O26

Statistical characteristics of races

Race		Canonical variables	
		Factor 1	Factor 2
Syriaca		1.826	3.569
Adami		4.067	1.842
Anatoliaca	− 0.109	− 0.789	
Meda		2.268	0.100
Cypria		− 0.317	2.637
Caucasica	2.965	− 3.779	
Armeniaca	− 0.387	− 0.327	

11.3 Characteristics and Natural History of Local Races

11.3.1 *Apis mellifera anatoliaca* Maa (1953):599

Introduction and History of Classification

Bodenheimer (1941) was the first to try a taxonomic classification of the bees of Anatolia, based on morphometric data using the characters given by Alpatov (Chap. 5). He was able to discriminate seven zones with different types: *A.m.caucasica* Gorb. in the NE, the "yellow Trans-Caucasian bee", referred to as "*A.m.remipes* Gerstäcker" in the region of Elazig (corresponding to *A.m.meda* in this study) and the "Central-Anatolian bee" in the central parts of the highland. This race is described as closely resembling "*A.m.remipes*" which "represents the true Irano-Turanian honeybee, with an area probably extending far into Central Asia". This was a prophetic sentence at that time!

Interesting is the description of the bees of West Turkey (Istanbul-Bursa) as differing from other types and "fluctuating between the Italian and the Syrian bee. No classification ought to be attempted as long as no analysis of the honeybees of the Balkan peninsula and of the Carniolan bee is available" (Bodenheimer 1941, p.29). The remaining three types are described respectively as intermediates between the Anatolian bee and Caucasians, yellow Transcaucasians and Syrians. It has to be noted that our own classification essentially matches this first attempt by Bodenheimer (Fig. 11.3).

In the formal taxonomic classification, published in 1953 and based on three museum specimens, Maa describes *A.m.anatoliaca*, subsp. *nova*, as follows: "Worker – Scutellum black. Abdominal terga black, the II anteriorly with a short, broad, pale band at each side. Tibia and basitarsus III exceptionally broad in profile" (there follow notes on scopal bristles, color of hair, ocellar distance). "Length of forewing about 10 mm. Queen and male: both unknown. Three specimens examined, from 'Turkey'. When further material be available, it may eventually prove to be a distinct species" (p.599).

Br. Adam (1983), who classified living colonies and bees by appearance and behavior reached essentially the same conclusions as Bodenheimer: four types in the W, NE, SE, and Central Anatolia and many intermediates. He collected numerous samples during three extended journeys across Asia Minor (1954, 1962, 1972). This collection, which was supplemented by Fuad Balci with samples from the extreme east and by A. Settar from the West of Anatolia furnished the basis of the present analysis. The invaluable importance of these samples is derived from the *time* of collection: before the period of general modernization (including migration of colonies) of apiculture in Turkey. In 1937, 112,000 traditional and 1300 modern hives were counted there (Bodenheimer 1941). In 1979, this number had increased to two million bee colonies, half of them in modern (Langstroth) hives, and 30,000 beekeepers practised migratory beekeeping over large distances. That is an almost 20-fold increase in total number and an 800-fold increase in modern hives (Inci 1980). A general hybridization took place and these early collections, therefore, became irreplaceable.

As shown on the map (Fig. 11.3), all of Turkey is covered by *A. m. anatoliaca* (including European Turkey), except for the NE (*A. m. caucasica)* and the SE (*A. m. meda*). In all multi-variate analyses, performed either with its immediate neighbors or with the bees of the Balkan, *A. m. anatoliaca* appears as a solid cluster. There is, however, one peculiarity: the bees of West Anatolia (west of the line Istanbul-Bursa-Eskeshir-Isparta) are represented as a coherent group at the periphery of the cluster. The morphological differences, compared to the samples of Central Anatolia, are shown in Table 11.2. In spite of the compact, permanent shape of the cluster "West Anatolia", I hesitate to recognize it as separate taxon, on account of its closeness to the general *anatoliaca* cluster; it seems preferable to treat it as a local population of this subspecies. Its special importance is based on the close ties to the bees of the East Aegean Islands (thus forming an "aegean race") and of Crete (*A. m. adami*).

Br. Adam stressed the difference between the bees of the Pontic coast (Sinop) and of the Mediterranean coast (Mersin) of Anatolia; the latter are said to show traits of the Syrian bee. In our analysis no clear separation is found between north and south, though some indications (with much overlapping) can be seen.

One fact, however, is an important result of the present analysis: the yellow bee from the NE corner of the Mediterranean, from SE Anatolia, is *not syriaca* as generally assumed previously, but the Iranian bee, *A. m. meda*.

As to the description of the exterior of the Anatolian bee, it seems appropriate to compare it with the Italian bee, which is similar in size and color and, moreover, is known all over the world. *Anatoliaca* is a little larger in all dimensions, with slight modifications in proportions: abdomen and tarsi are broader, the fore wing more slender, the legs and wings shorter in relation to body size. The *anatoliaca*, in general is a "yellow bee", too, but the rings on the abdomen are rather a smudgy orange turning to brown on the posterior segments (Br. Adam 1983). The scutellum is dull, dark orange. In wing venation there are characteristic differences, especially in a number of angles and in a much lower cubital index. Drones (n = 5 samples) are of similar size to *ligustica* but darker (including scutellum) with low CI ($\bar{x} = 1.575$)

Table 11.2 *Apis mellifera anatoliaca* compared to the population of West Anatolia (see Fig. 11.3) Mean and standard deviation (sd).; measurements in mm

Geographic type	n	Terg 3+4	F Wing length	Hind leg	Btars Index	Hair L	Toment I	Stern6 Ind	Color T2
A. m. anatoliaca	40	4.46	9.19	8.10	57.6	0.29	2.13	82.6	5.36
sd		0.09	0.13	0.13	1.0	0.03	0.26	1.9	1.43
Population "West Anatolia"	8	4.47	9.07	8.13	57.0	0.31	2.37	84.3	4.71
sd		0.12	0.15	0.8.13	57.0	0.03	0.42	3.8	2.02

Br. Adam (1983) studied the behavior and performance of the Anatolian bees and the hybrids during his travels and in his apiaries for well over 30 years. He especially stresses the adjustment to the extreme conditions of the upland steppes of Central Anatolia: energetic food-collecting activity, prompt reaction to periods of dearth by reduction of brood-rearing activity to save reserves and energy and a wintering ability exceeding that of races of northern origin (e.g., during severe winters in England). It shares to some extent the ample use of propolis with *caucasica*, but not its gentleness (which is noted on the eastern part of the Black Sea coast, being influenced by *caucasica*). Drifting of bees is low, especially if compared with *ligustica*: only half of the queens are lost during the mating flight (compared to other races). On the negative side is the special sensitivity to bee paralysis and the difficulty in processing the special nectar of *Calluna* properly.

The severe conditions for survival of bee colonies in Central Anatolia are impressively demonstrated in the study by Bodenheimer. During the years 1933–1939 censuses of beehives were taken by government officials all over Turkey. In bad years, in most of the districts more than half of the colonies were lost, in some of them even 90%. In wet years a colony could yield 10–12 swarms, resulting in an enormous fluctuation in the number of colonies. The general reliability of the data was verified in some cases by personal visits by the author. Moreover, in the districts along the Mediterranean coast no such fluctuation was registered. Therfore, it can be taken as proven that a permanent strong selection pressure was exerted on the bee colonies. At this time, apicultural activities in Turkey were restricted to honey harvesting from small pipe hives (volume 29l) and capturing of swarms.

Apiculture has an old tradition in Anatolia. In the Hittite code of Bogasköy, dated about 1300 B.C, several paragraphs are on beekeeping and they refer to an even older Codex. The value of a given weight unit of honey is mentioned as equal to the same unit of butter, and the price of a bee hive was the same as that of a sheep.

11.3.2 *Apis mellifera adami* Ruttner (1975a):271 (Origin of name: Brother Adam, Buckfast Abbey)

The morphometric analysis of four samples of worker bees collected by Br. Adam in 1952 initiated a more thorough study of the bees of Crete (Ruttner 1975, 1976a, 1980a). Samples of worker bees and drones were collected from different parts of the island as well as from islands of the Aegean Sea, in order to compare them with the bees of the Balkan Peninsula and of Asia Minor. To our great surprise, similarity was found not to bees of Greece (as was to be expected), but to the bees of West Anatolia and of the Mediterranean east coast (*A. m. syriaca*).

The statistical distance (d for *anatoliaca-meda* taken as 100, see p. 181) was 132 for *adami-cecropia*, about 75 *adami-anatoliaca* (population West Anatolia), and 66.6 *adami-syriaca* (Fig. 11.2).

In a PC analysis of the whole eastern Mediterranean two groups of clusters are represented (Fig. 11.1):

1. *Cypria-syriaca-anatoliaca*(west)-Aegean islands-*adami* (= oriental group)
2. *Intermissa-sicula-carnica* (Ionian islands) (= central group)

Description. a) workers. A. m. adami is large, within the oriental group second in size only to *A. m. caucasica*. Remarkably, it is larger (and darker) than *ligustica*, although living at about 10 degrees latitude farther south (35° N to 45° in northern Italy). Compared to other races there are several characteristic proportions: shorter and also narrower wings than *ligustica*; in consequence, a larger body has to be carried by smaller wings. Abdomen broad in contrast to other races of the region (index st6 81.2). Exceptionally wide distance between the two wax plates of st 3 (Fig. 6.6): while this distance generally fluctuates around 0.30, it is 0.392 mm in *adami*.

Cover hair of medium length, tomenta very broad (index 3.05). Pigmentation of tergites very variable, even among individuals of the same colony; some have broad yellow bands on t3 like other SE races, some are dark. The medium value of this tergite (5.6) for all samples indicates a narrow yellow band. The scutellum, however, is uniformly dark. Cubital index (very low) and other characteristics of wing venation differ substantially from neighboring races.

b) The drones. (n = 10 samples) are found in several scattered groups in the central region of a factor analysis, close to that of *anatoliaca*.

They are below medium size compared to *intermissa* and distinctly smaller than other races of the region (*anatoliaca, sicula, ligustica*). In contrast to workers, drones are uniformly dark (including scutellum). The CI is low ($\bar{x} = 1.39$).

Summarizing, *A. m. adami* is a very well-characterized island race (population in 1976 about 75,000 colonies). It is somewhat difficult, however, to delimit the border of this population to the east. The bees of the East Aegean Islands (Karpathos, Rhodos, Kos, Chios, Lesbos) are very similar to *A. m. adami* (Ruttner 1980a). Further, only a gradual variation is found among the bees of these islands and of the west of continental Asia Minor. *Adami* and *anatoliaca* doubtless constitute well-defined, distinct races. The taxonomic classification of the bees of the Aegean Islands, taking an intermediate position, seems to be a question of the individual view of the taxonomist. Subspecies groups frequently show gradual variation, and in these cases it may be appropriate to take the situation as testimony of local evolution rather than to impose rigid classifications.

Behavior. A. m. adami winters without difficulties also in cold-temperate climate (Br. Adam 1983). The brood-rearing activity of the colonies is astonishing: brood rearing continues during winter due to the favorable climate of the island. Beginning with February, a sharp increase of the brood quantity starts, up to a peak of 14–18 Langstroth frames in May. At this point, almost every colony prepares to swarm, constructing usually 60–200 queen cells, partly in clusters (Ruttner 1980, Br. Adam 1983.)

Br. Adam describes this bee as "extremely aggressive", especially when kept in the cool climate of England. In Crete, the defense reaction was found to be quite variable; under favorable conditions colonies can be opened without any protective clothing, on other occasions the bees attack as soon as a hive is approached. The behavior of the bees on the comb is frequently very calm, they usually remain quiet even if removed from the hive. The morphometric distance to the bees of Greece and the affinity to races of the East Mediterranean gives rise to an interesting problem of zoogeography. The Aegean region had an unsteady geologic history during Tertiary and Pleistocene, changing between elevation and depression beneath sea level. The island of Crete, however, always had solid soil, at least in part (Kuss 1973, 1975). The islands of the Aegean show faunal influences from Greece as well as from Anatolia. In Crete, most of the vertebrate fauna of the Pleistocene came from the Peloponnes. The "Karpathos trench" between Karpathos and Rhodos was an important divide between the fauna of Europe and Asia Minor. However, there are also examples of common elements with Asia. Two species of rodents (*Apodemus* and *Acomys*) and the Bezoar goat *Caprus aegagrus* occur both in Anatolia and Crete, evidently having crossed the Karpathos trench. *A. m. adami* provides another example of the ponto-aegean distribution of a taxon.

Since prehistoric times Crete has been "bee country" (Ruttner 1976 a) and this is true today. The island is one of the regions with the highest density of colonies: in 1976 the total population amounted to 75,000 colonies, that is an average density of ten hives per square kilometer. The nectar source is mainly the native Mediterranean flora. Even on barren mountain slopes (especially in the south) where agriculture is reduced to a few sheep and some poor olive trees, bees collect a fair honey crop from wild thyme growing in the rock crevices. Therefore apiculture is an important economic factor for the population (Zymbragoudakis 1979; Ruttner 1976, 1980).

11.3.3 Apis mellifera cypria Pollmann 1879:25

The Cyprian bee represents the island race with the smallest territory. Nevertheless, it early received great publicity among apiculturists, in the first place probably for its "exotic look". In his book on bee races, A. Pollmann (1889) devoted more than 17 pages (out of 70) to this bee. "This bee is a true apistic beauty" (Cori in Pollmann 1889).

As we have seen earlier (Fig. 11.1) *cypria* shows similarities to the neighboring races, corresponding to its geographic location. Statistically, it is almost equidistant from *anatoliaca* (74.6 units), *syriaca* (69.8), and *meda* (87.3). The distance to the centroid of the *meda* population "SE Anatolia", however, amounts to less than half of this value.

If compared with *anatoliaca*, there are specific differences: *cypria* is smaller in all body dimensions, but relative to body size it is long-legged and long-tongued, with relatively shorter wings and the most slender abdomen of all oriental races (Index St6 84.6). The cubital index is high (2.72). The most conspicuous difference, noted by all observers, is color: the yellow zones on the abdomen and the scutellum are not only larger than in *anatoliaca* (Table 11.1), the color is also brighter – a shining carrot-orange – giving the bee the esthetic appeal so much appreciated in the apicultural world 100 years ago. The drones, too, are described as having orange rings on the abdomen (Buttel-Reepen 1906; no *cypria* drones were examined in this study). Similar color pattern and slenderness was the reason that the Cyprian bee (as well as the Syrian) was believed to be related to the Egyptian (Buttel-Reepen 1906; Br. Adam 1983). As shown by morphometric analysis, however, *A. m. cypria* belongs to the oriental group of races, close to its genetic center, *anatoliaca*, while *lamarckii* is a member of the "Tropical-Africa" group.

The behavior of the Cyprians as described by Buttel-Reepen (1906) and Br. Adam (1983) points to the relationship to the bees of Anatolia. There is the surprising fact that *A. m. cypria*, native to a country with subtropical climate, "surpasses in hardiness and wintering ability every other variety of the honeybees" (in England; Br. Adam 1983). Many queens were shipped to America and Europe from Cyprus since 1880 by F. Benton and his successors (Buttel-Reepen 1906). Not wintering problems, but the energetic defense reaction when disturbed finally prevented this bee from becoming established in other countries, as did the Italian bee. Even smoke does not calm it – it reacts even more fiercely.

All the authors mentioned above emphasize (together with further details on behavior) the "amazing fecundity" of *cypria* during the whole season until autumn, and much of the honey stored is converted into brood (in this respect differing from other Mediterranean races showing a bimodal distribution of brood amount during summer). In preparation for swarming, up to 40 queen cells are constructed, frequently several close together.

The total population of *A. m. cypria* is relatively small. According to Br. Adam (1983), about 22,000 colonies were counted in 1952 on the island of Cyprus (plus, undoubtedly, a certain number of feral colonies). In consequence of the introduction of modern apiculture, the mumber of colonies increased from 23,500 in 1965 to about 50,000 in 1972 (Church and White 1980). Contrary to expectations, a considerable variability in morphology is observed.

11.3.4 *Apis mellifera syriaca* Buttel-Reepen 1906:175

The Syrian bee autochthonuous on the mountains and in the valleys east of the Mediterranean (Israel, Jordan, Lebanon, Syria) is the smallest of all oriental and Eropean races of *mellifera*, as stated by Bodenheimer (1941). This was confirmed by the present analysis (Table 11.1): only African races (including *yemenitica*) are smaller. The small size of the bee is expressed by the cell pattern of the comb: 484 worker cells are found per dm^2 versus 427 in European colonies (Blum 1956).

In comparison with *A. m. cypria*, tongue, hind wing, and abdominal terga are considerably shorter, but the hind leg only slightly so; this means this *syriaca* is a relatively long-legged bee. The tibia is even longer in terms of absolute units than that of *cypria*. The basitarsal index is lower than in most oriental *mellifera* bees, approaching *ligustica*. Abdomen slender; cover hair very short; color of terga and scutellum bright yellow; cubital index low. In all graphic representations of multivariate analyses the fairly compact cluster of *A. m. syriaca* is found in a peripheric position in regard to the oriental group, located in the direction of the African races *(lamarckii, yemenitica).*

Syriaca was early introduced to the apicultural world by F. Benton, who shipped queens to the USA and England, and by the well-known apiculturist Baldensperger, who managed a professional apiary in Palestine for many years (Buttel-Reepen 1906). Since these bees are very sensitive to cold, as well in their activities during summer as during winter, these importations to cooler zones were not successful (Br. Adam 1983). In their native country, however, they are excellently adapted. They produce more honey than imported Italians and show a more effective defense strategy against the most powerful bee predator in this region, *Vespa orientalis,* by stopping any flight activity during the hot hours in the "hornet season" (autumn). Even so, this wasp may cause problems for *syriaca* too, since it blocks the flight activity of the bees to a great extent (Blum 1956, Kalman 1973).

Although *syriaca* is a good nectar collector, colony management is severely hampered by fierce defense reactions. The intruder is followed over 500 m and if the bees are alerted, animals (horses, camels, mules) may be attacked and killed. Several other behavioral charateristics are found in *A. m. syriaca* besides this easily alerted, fierce defense reaction and the adaptation to a permanently warm climate as described above: little use of propolis, construction of very many swarm cells – up to 200 or 300 – and survival of dozens of young queens in the colony until the first of them is mated and laying eggs. This important behavior is observed also in *A. m. lamarckii, A. m. intermissa* and *A. m. sicula (see relevant Chapter 12, 13, and 14).*

The phenomenon of relinquishing the habit of strict monogyny during the period of reproduction in some races of A. mellifera deserves some discussion, because this seems to be a true enigma in the evolution of this species. In view of the fact that a number of races succeeded in solving the problem of queen losses during mating by breaking the principle of strict monogyny for a short period, we have to ask the question: why did other races not adopt this system, which seems to provide a selective advantage of about 20 %?

In all other races more than one queen is tolerated in the colony only exceptionally. When an afterswarm leaves the colony, several virgins may hatch simultaneously and join the swarm. They are eliminated, however, in the swarm cluster formed at an intermediate site. If the single remaining queen is lost on the mating flight, the whole swarm is condemned to perish. The same may happen to the mother colony, because at the time when its virgin queen starts for the mating flight no young larvae are left to replace her.

If, however, several virgin queens remain in the afterswarm and in the mother colony, a lost queen can immediately be replaced by another virgin. Since surplus virgins remain available until one queen is ready for oviposition, virtually no queenlessness whatsoever would occur. In European races about 20% of the queens, on the average, (and in consequence, their colonies), are lost during the mating flights. To avoid these losses would imply an enormous selection advantage. How can it be understood that this behavior did not spread to other races? Could its isolated occurrence be interpreted as indication of ongoing evolution?

The results of morphometric analysis indicate that in the year of collecting the samples (1952) even in Israel mainly pure *syriaca* bees were present. Beginning with 1950, an official program was started in this country to replace the native bee by more gentle Italian stock from the USA to enhance modern apiculture. By 1958, already 80% of the 44,000 colonies registered at this time were of Italian race (Ben Neriah 1958). In Jordan, however, even in 1981 about 85% of the 40,000 colonies were kept in traditional mud pipes (similar to Egypt) and only a few Italians had been imported (Robinson 1981).

Several questions remain unanswered concerning *A.m.syriaca*. Between this race and *A.m.anatoliaca* a narrow area of *A.m.meda*, population of SE Anatolia, is squeezed in, in the NE corner of the Mediterranean (Antakya; Fig. 11.3). Is there a direct contact with *syriaca,* and are there intermediates? The same problem exists in Syria: how far to the east does *syriaca* extend, and is there any borderline to the population of *A.m.meda* in Iraq?

11.3.5 Apis mellifera meda Skorikov (1929b):261

This name was given by A.S.Skorikov to a bee occurring in the USSR close to the Iranian border and differing from the "yellow transcaucasian bee". Only tongue length and shape of the abdominal sternites (a characteristic no longer used in taxonomy) were recorded as description of this race. "Distribution: North Persia. Lencoran" (located on the Caspian Sea). The same bee and its distribution was described in detail by Ruttner et al. (1985b).

Multivariate analysis from data of 63 samples from Iran together with samples from Turkey, Syria, Lebanon, Israel, Jordan, Cyprus, and Iraq resulted in various clusters, one of them including all samples from Iran, plus all samples from Iraq (collected in the northern mountainous part which is the only one of this country with native bees) and from a restricted area in SE Anatolia. Since the cluster includes also the samples of Azerbaijan, close to the Soviet border, it represents the same *A.m.meda* first described by Skorikov. It is the bee of the Iranian highland west of the big deserts Kavir and Lut, extending with a relatively narrow zone as far west as the Mediterranean (Fig. 11.3). The coast of the Persian Gulf is void of *A.mellifera* (see Chap.7, *A.florea*).

As far as the statistical distances to other races are concerned, the distance between the centroids *meda-anatoliaca* was used as standard ($=100$) for all analyses of the oriental *mellifera* region. Only the centroid of *cypria* is closer to *meda* than the standard, indicating the peripheric, rather isolated position in relation to the other (87.3) races.

Description. a) Workers. This paragraph can be very short: the exterior of *A.m.meda* resembles in an astonishing way that of *A.m.ligustica* (Table 11.1; a strange parallel evolution!). The *ligustica* cluster is regularly completely dispersed if samples of both races are analyzed together with others in a regional PCA. In the collective PCA of all *mellifera* races (Figs. 10.5, 10.7) the *meda* cluster is situated in a central position, the *ligustica* cluster in the C-branch. By special discriminant analysis a clear canonical separation is achieved, however, based on five characters (two characters of size – length of tergite 3 and width of fore wing, length of cover hair, and the angles G18 and I10 of wing venation).

Since the two races belong to different geographic groups, three additional differences in complex characteristics not used in statistical analysis (Table 11.1) were found: *meda* has broader metatarsi, a broader abdomen and slightly narrower fore wings. Moreover, the scutellum is bright yellow (class $\bar{x} = 5.95$) and not predominantly dark (class $\bar{x} = 1.95$) as in *ligustica*.

b) Drones. Almost the same as for workers can be said: *meda* drones are similar to *ligustica* drones, with only slight differences. Length of fore wing and hind leg slightly shorter, cubital index lower (1.89).

Pigmentation on the abdomen is darker than in workers, with only one yellow band (class 5.3) on T2. Scutellum variable, mostly only slightly brownish in both races. Basitarsal index, elevated in workers, is low in *meda* drones (46.42 – thus not "remipes"), even inferior to *ligustica* drones (47.95). In a factor analysis drones (n = 20 samples) are represented in a compact cluster in a peripheric position (Fig. 10.8). Variability in size is considerable although only drones from Iran (and none from Iraq or Turkey) were available. Considering the total range of size variability in *A. mellifera*, these drones take a medium position (Ruttner et al. 1985). Drones from the SE (Kerman) and the Caspian Sea tend to be the largest (as are the workers).

Six local populations can be discriminated within *A.m.meda*, with partly overlapping variability (Figs. 11.3, 11.5; Ruttner et al. 1985):

1. West and central Iran (Azarbaijan – Iranian highlands), average altitude 1200–1500 m with mountains higher than 4000 m and bee colonies up to 3000 m.

2. Subtropical coast of the Caspian Sea (Mazandaran)

3. NE Iran (Mashad) represented by an especially well separated cluster

4. SE Iran (Kerman)

5. Iraq

6. SE Anatolia, from Van Lake to Antakya. This represents the population with the smallest bees, but with long proboscis, coming close to the cluster of *A.m.cypria*. A similarity with this race is also found in the metatarsi, which are even broader in SE Anatolia (index 59.04). It is interesting to note a gradual decline of the basitarsal index from west to east – to values as low as 54.56

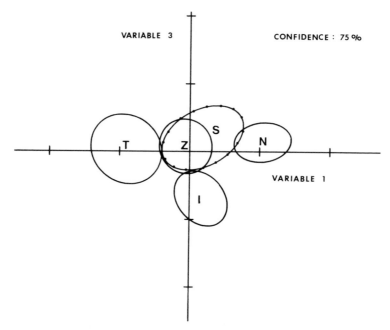

Fig. 11.5 Local populations of *A.m.meda.* Discriminant analysis (n = 63), ellipses of 75% confidence. *I* Iraq, *N* NE Iran, *S* S Iran, *T* SE Anatolia, *Z* Central Iran

in NE Iran (Mashad). This bee was wrongly classified as "*A.m.remipes* Gerstäcker" by Bodenheimer 1941.

The high variability in altitude and climate within the area of distribution of *A.m.meda* provides an occasion to test whether general rules of zoogeography (investigated mainly in vertebrates) apply also to honeybees. Somehow, only the population with smaller bees inhabiting lower altitudes in the mountains of Iraq conforms to Bergmann's rule. In the most extreme biotopes - Mazandaran situated in a depression with humid, subtropical climate and the southern semidesertic region of Kerman (alt. 1700 m) with hot summers (max. 40°) - the populations with the largest bees occur; this clearly contradicts this rule.

There is another puzzling situation. In the NE corner of the Mediterranean (area of Antakya) three populations with distinct characteristics meet: *A.m.anatoliaca, A.m.meda* (population SE Anatolia), and *A.m.syriaca.* Climatic differences are found *within* the area of the race (coast-mountains), but, in this special location, not *between* races. Observations of this kind have to be mentioned to avoid undue generalizations.

Physiology and Behavior. A.m.meda is one of the races of *A.mellifera* which, according to present knowledge, have never been transplanted. However, an adaptation to long winters can be concluded from the meteorological data, e.g., from the Zagros mountains (alt. of bee site 2600 m), showing days with frost for 6 months of the year. There exists a strong swarming tendency, but in contrast to *syriaca*, the number of queen cells is only mediocre (10-20). Much propolis is used. The bees are more easily alerted and prone to pursue an intruder (for up to 200 m) in the south (Iraq) than in the north (Azerbaijan).

A.m.meda is one of the "big" races, with a large territory and more than one million colonies in Iran alone, and with beginning taxonomic radiation.

11.3.6 Apis mellifera caucasica Gorbachev (1916):39

The "grey Caucasian mountain bee" is one of the four honeybee races known all over the world and has been used in beekeeping for more than 100 years (Pollmann 1889). The general morphological, behavioral, and economic characteristics are universally recognized. This bee was extensively investigated by Soviet authors, but on reviewing the bibliography the scientific classification is found to be far from clear. This is due to a very confusing history of nomenclature. To start the discussion from an undisputed point: there is unanimity that the "Caucasian mountain bee" occurs only on the main chain and in the southern valleys of the Caucasus and in the higher reaches of the Maly Kavkaz (Little Caucasus). This includes the east coast of the Black Sea, all of the SSR Gruziya, and parts of Azerbaydzhan (Alpatov 1948; Bilash et al. 1976; Avetisyan 1978).

The first honeybee from the Caucasus region was described (but never published, p. 145) as *Apis remipes* by Pallas in 1773, probably originating from Mozdok north of the Caucasus (Skorikov 1929). The name "remipes" was retained as subspecies (or even species) by Gerstäcker (1862); Buttel-Reepen (1906); Skorikov (1929 a); Maa 1953. Pollmann (1889) described colonies of bees imported from the caucasus region to Germany as "*A.m.caucasia*" but the exact origin of these bees is not known and the name "*caucasia*" was never used later. In 1916, finally, the "grey mountain bee" was scientifically described by Gorbachev as "*A.m.caucasica*" and this name was almost generally adopted in apiculture. To make the situation still more confusing, Alpatov (1929, 1948) presented a map showing the distribution of the grey mountain bee ("*A.m.caucasica* Gorb.") in the Caucasus and the "yellow transcaucasian bee, *A.m.remipes* Gerstäcker" in Armenia. This name was used also by Bodenheimer (1941) for the yellow bees of E Anatolia and Iran. On the other hand, all bees of the Caucasus were given the name "*Apis remipes*" by Skorikov (1929), with the subspecies *transcaucasica* and *armeniaca*.

From this short review and the morphometric analysis below it becomes evident that the name *A.m.caucasica* Gorbachev is most appropriate for the "grey Caucasian mountain bee" and that the yellow Armenian bee should be *A.m.armeniaca* Skorikov (1929) since no species "*Apis remipes*" is recognized. The hitherto pending question of taxonomic classification of the honeybees of the Caucasus is finally approaching a solution.

Distribution

Biometric data on the western and southwestern part of the Caucasus are available from Abchazia and Grusinia. A map published by Gorbachev (1916), which was referred to also by Skorikov (1929) and Alpatov (1929), shows the *caucasica* area as far as the east end of the mountains, that is across a distance of about 900 km. To the SW, *caucasica*-like bees are found on the Black Sea coast of Anatolia as far as Samsun, about 600 km distant from the Caucasus summits. In the lowlands, evidently, mainly hybrids between *caucasica* and *armeniaca* are found.

The climate of the Caucasian region is humid-subtropical at sea level and warm-temperate or cool-temperate in the mountainous and alpine zones. Lush forests are abundant, especially in the west, with many feral bee colonies which are crossed with managed colonies to form one single population (Skorikov 1929). Since the climate is similar in the long narrow *caucasica* branch along the Pontic coast of Anatolia, the *caucasica* area seems to be mainly limited by climate.

Morphology

The general aspect of *A.m.caucasica* is very similar to *carnica* (Fig. 11.6): about the same size, predominantly dark pigmentation (but narrow, frequently indistinct yellow bands on abdominal terga, Fig. 11.7), broad grey tomenta and short cover hair, somewhat broader abdomen (St 6-I 82,0 vs. 83,5). Basic differences are found in wing venation: the generally known cubital index is much lower than in *A.m.carnica* (Fig. 11.7). Even more than in carniolans, however, "dark color" in *caucasica* is rather a model for selection in beekeeping than reality: both races show high variability from an almost completely dark to a narrow yellow stripe.

One character of color seems to be uniform and specific in *caucasica*: the

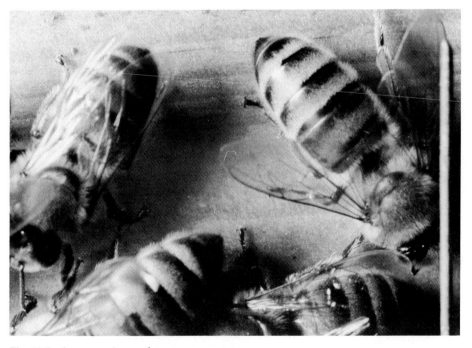

Fig. 11.6 *A.m.caucasica,* worker

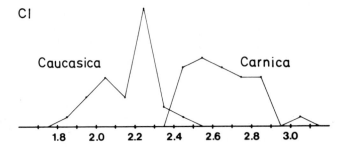

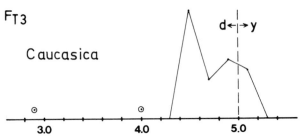

Fig. 11.7 Frequency distribution of cubital index of *caucasica* and *carnica* (top). Color classes in *caucasica* (bottom)

dense hair cover of the thorax of drones is deep black, different from all other known races. Hair color of workers is similar to *carnica*, but with a slightly different tinge; it is rather lead-grey compared with the brownish-grey of carniolans. The world-wide reputation of *caucasica* comes from the long proboscis. Biometrics in bees started with measuring tongue length, and Cochlov (1916), Michailov (1926), Alpatov (1927) and Skorikov (1929) stated that caucasian bees have the longest tongues of all *mellifera* races. The highest colony average was found to be 7.22 mm in Megrelia, with individual tongue length up to 7.52 mm (Skorikov 1929). In the present study the highest colony mean was 7.25 mm and the longest individual tongue 7.38 mm. Biometric data, compared to other oriental races, are given in Table 11.1.

Except for the race-specific characteristics described, there are only slight differences: compared to *carnica*, the *caucasica* has a slightly larger body and longer hind legs, but smaller wings; the cover hair is slightly longer, tomenta and the stripes of yellow pigmentation at abdominal tergites 2 and 3 are definitely broader. The most important differences between *carnica* and *caucasica* concern wing venation. In fact, *caucasica* shows a completely different venation pattern. Of 11 angles measured only two fall within the limits of the standard deviation while the differences of the other nine are all beyond these limits (total N = 959 bees). This is the reason for the special location of the *caucasica* cluster in the "Y" of the total PCA (factors 1 and 2; Fig. 10.5): differing from the expectation derived from the general aspect not close to *carnica*, but in the M-branch close to *mellifera*. In a graphic

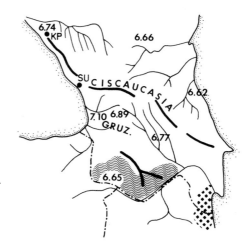

Fig. 11.8 Map of the Caucasus region. Numbers: length of proboscic (after Skorikov and personal data), *GRUZ* Gruziya, *KP* Krasnaja Poljana, *SU* Suchumi. Dots: *A.m.meda*, rolling lines: *armeniaca* (Skorikov 1929)

representation of factors 2 and 3 (with factor 1, containing characters of size, removed), *caucasica* appears at the end of a separate branch (0-branch) with all the *mellifera* races of the Near East (Figs. 10.6, 10.7). *Caucasica* drones are larger than all other drones (Table 11.4). Pigmentation of tergites is dark, the CI is even lower than in *mellifera*; basitarsi are broad compared to European races (as in worker bees).

To summarize, *A.m.carnica* and *A.m.caucasica* differ in many characteristics, but they are hard to discriminate without biometric and biologic data. No wonder that caucasians offered for sale are hardly ever pure but are mostly hybrids with carniolans.

Local Varieties

Investigating only tongue length, Skorikov (1929a) established four local populations: Abchasian, Megrelian, Georgian, and Imeretian (Fig. 11.8). Although the number of samples from the three last-named regions is small in our collection, the multivariate statistical analysis based on 40 characters seems to confirm Skorikov's results (Fig. 11.9): the Georgian samples (region Kutaisi) are well separated from the Abchasian in all plots of the first three factors, and in the Imeretian samples in two factors. The Megrelian bees have the longest proboscis ($\bar{x} = 7.20$ mm), the Imeretian the shortest ($\bar{x} = 6.60$), confirming the results of earlier authors. It was maintained that in the Caucasus, bees at higher altitudes are larger and darker than those of the plains (Skorikov 1929) corresponding to observations in other regions (p.55). Comparison of 16 colonies from Krasnaja Poljana situated between 100 m and 1800 m, showed no linear gradient in body size, hair length, tomentum width, tongue length, and color; but if bees from the lowland (100 to 500 m) and the mountains (600 to 1800 m) are compared (Table 11.3, first and last line), bees of the mountains appear to be slightly larger, darker, and with longer and more hair. Populations from different altitudes of one and the same region can be distinguished by multivariate analysis (Fig. 11.9).

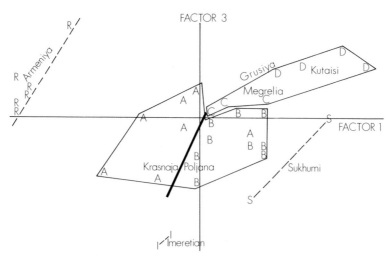

Fig. 11.9 Local populations of the Caucasus region: PCA of 36 samples from various location (factor 1+3). Within the cluster Krasnaja Poljana a gradation according to altitude from left to right; the oblique line divides lowland (100 to 500 m) from mountain (600 to 1800 m) samples

Table 11.3 Characters of *A.m.caucasica* in different altitudes of the region Krasnaja Poljana (N = 16 samples). Mean and standard dev.(in paranth.). Terg. = sum of longit. diameter of tergite 3+4; Hair = cover hair; Tom. Ind. = tomentum index. Color T3 = pigmentation of tergite 3.

Altitude(m)	Terg. 3+4	Probosc.	Hair	Tom. Ind.	Color T3
100–500	4.50	6.970	0.330	2.37	4.70
(n=4)	(0.07)	(0.122)	(0.007)	(0.18)	(0.15)
600–1150	4.54	7.057	0.337	3.09	4.63
(n=3)	(0.04)	(0.031)	(0.009)	(0.40)	(0.26)
1200–1600	4.55	7.081	0.347	2.86	4.64
(n=6)	(0.03)	(0.058)	(0.009)	(0.14)	(0.21)
1650–1800	4.52	7.053	0.328	3.05	4.52
(n=3)	(0.01)	(0.050)	(0.002)	(0.17)	(0.20)
600–1800	4.54	7.068	0.340	2.97	4.48
(n=12)	(0.06)	(0.035)	(0.015)	(0.38)	(0.15)

Biology

A.m.caucasica is a race of general economic value, and many comparative tests have been made in apicultural institutes to evaluate the relative performance under different climatic conditions (Alpatov 1932, 1948; Avetisyan 1978; Bilash 1967; Bilash et al. 1976; Winogradowa 1976; and others). There is agreement on quite a number of points:

1. Very high level of gentleness
2. Low swarming tendency. Modest number of swarm cells (10–20)
3. Flat curve of brood-rearing activity: late and slow start in spring. Medium population size in summer and autumn.
4. Flower preference: colonies excel in a long protracted slight nectar yield; much

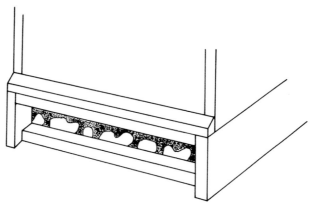

Fig. 11.10 Propolis curtain at the entrance of a *caucasica* colony

less so in short mass offer of nectar, which is likely to be stored in the brood chamber and not in the super. This reflects the warm humid climate with availability of nectar during a long season. More visits of red clover than other races (Alpatov 1929), although not all expectations were met in this regard (Goetze 1964). "Wet" capping of honey cells (no air between capping and honey).
5. Excessive use of propolis. During winter the major portion of the entrance is blocked by resin, leaving only small holes for ventilation and flight activity (Fig. 11.10).
6. Poor winter survival evidently due to higher *Nosema* sensitivity (according to several reports, e.g., Alpatov 1932; Avetisyan 1978). No shipping of caucasians to Siberia is permitted (Gajdar 1976).
7. Excellent combining ability with other races, especially with *ligustica* and *carnica*.

Collective efforts were made to select suitable bees for different climatic regions of the USSR (Bilash 1967). The caucasian mountain bee gave the best results in the Black Sea region (Krasnodar, Moldava). The population of *A. m. caucasica* is presumably rather high, due to generally favorable conditions. The number of managed colonies in the SSR Grusiya was given to be 175,000 (Andguladze 1971).

A special taxonomic problem are the bees of the north Caucasus ("Ciscaucasian bee"). Alpatov (1929, 1948) found that this bee continues the tendency of vari-

Table 11.4 Selected morphometric data of drones of the subspecies *caucasica*, *carnica* and *mellifera*. Measurements of length in mm.

Subspecies	N	F Wing length	width	H leg	Ct2	Venation angles B4	E9	I10	CI	BTarsI
Caucasica	4	12.41	4.05	10.25	1.10	104.7	18.8	62.4	1.38	50.5
s. d.		0.26	0.11	0.35	0.30	6.6	1.2	4.8	0.24	2.2
Carnica	21	12.18	3.99	9.87	0.85	114.5	22.1	56.1	1.93	48.1
s. d.		0.33	0.13	0.30	1.21	6.2	2.0	5.3	0.34	2.7
Mellifera	7	12.24	3.92	10.10	0.06	105.1	18.6	52.1	1.43	46.0
s. d.		0.51	0.20	0.27	0.27	7.0	1.9	4.3	0.27	3.6

ation of body measurements from north to south in the plains of the USSR (increasing tongue length, decreasing body size) (Table 4.5). This tendency is abruptly changed at the Caucasus mountain chain. The Ciscaucasian bee which since Gorbachev (1916) frequently has been regarded as a hybrid of the mountain bee was much less studied biometrically than the other populations of the region. Tongue length was given as 6.50–6.80 mm, smaller body, but longer legs compared with the Ukrainian bee, yellow stripe on tergite 3 (Alpatov 1929). At the present time the bees of this region are being replaced by carpathian strains.

11.3.7 *Apis mellifera armeniaca* Skorikov (1929b):262

The "yellow armenian bee" was first described biometrically by Alpatov (1929), but without attributing a scientific name. He found smaller body dimensions (especially a shorter proboscis), and a higher cubital index than in *A. m. caucasica*, and more yellow body pigmentation than in *ligustica*. Gorbachev (1916) presented a map of the distribution of the "yellow transcaucasian bee" (Figs. 11.3, 11.8), but he did not discriminate betwen the Persian bee (*A. m. meda*), in the area around Lenkoran on the Caspi Sea and the Armenian bee, which is quite different as shown by samples from a bee yard in the Megri region in SE Armenia in our collection (Table 11.1). The Armenian samples have the same body size, but distinctly longer tongue, shorter wings, longer hair, broader tomenta, a much brighter pigmentation, and higher CI. The characteristics given by Markosjan et al. (1976) for bees of the same region are in good agreement with these results.

In all multivariate analyses the Armenian samples are very close to the *anatoliaca* cluster, partly overlapping. Because of the few samples examined, the taxon *A. m. armeniaca* Skor. is retained only provisionally, considering the possibility that the bees of ASSR Armeniya are the easternmost population of *A. m. anatoliaca*.

The behavior of the Armenian bee is described by Awetisyan (1978) as aggressive, nervous on the comb, a great brood producer, not much inclined to swarm. Being susceptible to nosema disease and honeydew intoxication, its wintering ability in cool climate is allegedly poor. Nevertheless, in Armenia this bee occurs up to 1500 to 2000 m, surviving a winter of 5 months without difficulty (Markosjan et al. 1976).

11.3.8 Annex: *Apis mellifera taurica* Alpatov (1938b):473

V. V. Alpatov (1938b) collected several bee samples from basket hives owned by a tartar farmer which turned out to belong to an isolated race, the "Crimean bee". Morphometrically, this race is a member of the oriental group: broad metatarsi (index 57.73), large body size (only slightly smaller than *caucasica*), dark pigmented abdomen. The cubital index is below 2.0 (since Alpatov's method of measurement is different, his data are only indirectly comparable). This race was said to be very gentle and up to 100 queen cells are allegedly constructed during the swarming season. No further report since the first description is available to the author.

CHAPTER 12

Honeybees of Tropical Africa

12.1 General

12.1.1 Size of Bee Population

In Africa south of the Sahara probably more bee colonies are found than in the remaining part of the orginal *mellifera* area in Eurasia. Most of the colonies exist in feral condition, nevertheless the above statement is supported by an estimate of the number of bee colonies derived from the annual export of beeswax as given below (Table 12.1). These figures are available for only a few countries since in many others no export of beeswax takes place at all, in spite of a similar density of the bee population, because no market is available for this product.

The number of colonies in Ethiopia in 1975 was estimated to be 3 million and the quantity of exported wax was 350 tons (Gebreyesus 1976). This gives a very rough estimate of the relation between amount of wax harvested and the number of bee colonies.

12.2 Characteristics Common to Tropical Honeybees

A number of peculiarities are reported for all tropical countries of the continent, e.g., high swarming rate, rapid colony growth, various types of poorly protected nesting sites, tendency to abscond and to migrate, etc. Bees with these traits proved to be extremely successful also in the tropics of S America and quite similarly in the tropics of S Asia. Therefore, the behavior described in this section has to be regarded as tropic-specific for honeybees. To understand the adaptive signif-

Table 12.1 Beeswax export of different African countries. (Data from Dubois and Collart 1950; Douhet 1965; Gebreyesus 1976)

Year	Country	Metric tons
1947	Angola	1121
1947	Tanganyika	410
1947	West Africa (French speaking zone)	505
1947	Equatorian Afica	319
1965	Madagascar	1100
1976	Ethiopia	350

icance fully, the most important environmental conditions faced by the tropical bees have to be investigated.

12.2.1 Predation

The figures on the export of wax reveal yet another fact: the enormous number of colonies destroyed every year. In restricted parts of various countries, methods of true beekeeping are certainly practiced, without even touching the brood nest (Dubois and Collart 1950; Nunes and Tordo 1966; Lepissier 1968; Nightingale 1976; Svensson 1984). Mukwaira (1976) reported that hunters formerly depended on venison and honey for their livelihood on their hunting expeditions in S Africa; they spared exploited bee colonies to save them for future use. However, little wax is harvested by these methods and the most frequent and easy way to obtain honey and wax is to hang empty wicker or log hives on trees, hope for rapid colonization (which is likely to occur on 50-90% of the hives; (Drescher 1975; Guy 1976; N'Daye 1974) and to "harvest" the colony several months later. Usually all the combs are taken, including the brood, and although the bees are not always killed by intention, the hazard to the colony is great, since this is done during the night. Sometimes, however, all bees are killed prior to the cutting of combs. Drescher (1975) and Fletcher and Tribe (1977b) report an average of 0.5 to 1 kg of wax obtained per colony, a further indication that all the combs are removed.

Honeybees in Africa most likely did not exist earlier than the beginning of the Pleistocene (Chap.3). This means that they coexisted with early man from the beginning and that honey has always been an essential food. This is true today for the pygmies of the Congo and bushmen of the Kalahari. Heavy predation of the feral honeybee population by man is observed in every country of tropical Africa. The methods are rationalized: feral swarms searching for nesting sites are attracted by favorable cavities, housed for several months, and then destroyed; subsequently, this procedure is repeated. Defending against predation and its consequences was recognized to be a major factor for survival of honeybees in tropical Asia by Seeley et al. (1982). As in Asia with other *Apis* species, man is the most efficient predator of the bee population in Africa, with much greater impact than other predators, such as the honey badger *(Mellivora capensis)*, *Merops*, wasps *(Palarus* and *Philantus)*, and ants.

12.2.2 Strategies for Survival

Intense defense reaction, a well-known attribute of African bees, may deter dilettantish attempts at honey collecting, but it is no help against skilled hunters using smoke and fire. Survival in the presence of an invincible enemy can be secured in two ways: (a) by escape and (b) by an increased reproduction rate. Early abandonment of the nest in case of all kinds of disturbance (absconding) is found in all tropical honeybees, most in the Little honeybee, *A.florea*, and least in *A.dorsata*, with its powerful defense reactions. In the tropics, absconding is very likely the salvation if an attack cannot otherwise be stopped. Near Tanga (coast of Tanzania), *A.m. litorea* colonies first try to evade attacks of the wasp *Phalarus* by stopping flight activity. If attacks persist and increase, the colony will abscond (Smith 1961). In the temperate zone, absconding would be deadly except for a short period in the early season. Therefore, all tropical races are highly mobile, races of the temperate zone are stationary (equally in *A. cerana* as in *A. mellifera*).

An experience reported from Madagascar shows that this difference in behavior exists and is maintained even between two neighboring populations of the same race (*A. m. unicolor*). According to Douhet (1965), beekeeping can be practised on the highland of the island just as it is done in Europe: the bee is quiet and does not react unexpectedly after manipulation of the colony. When a school teacher and enthusiastic beekeeper was transferred from the temperate highland to the tropical lowland he soon became completely discouraged with the local bees: "no sooner had he opened the hive the bees absconded. He was not able to harvest honey without losing the bees" (the bees of the tropical coastal plain of Madagascar probably are a separate local ecotype, see p. 226).

Absconding colonies must have a chance to find a new site if they abandon their nest. Koeniger (1975) pointed to the adaptive advantage of open-air nesting in the tropics. *Mellifera* races there are observed to be much less choosy when selecting a new nesting site: besides nesting in hollow trees and clefts in rocks they also use cavities in the ground, and termite mounds, metal drums, the underside of any kind of roof, or telegraph poles (Anderson et al. 1973; Fletcher 1978a). Exposed sites are probably more dangerous, but they may be better than nothing. A peculiar story is reported from S Africa. In the swampy ground of Springbok Flats, Letaba District, NE Transvaal, clefts and cavities arise in the dried mud during the winter period which are quickly occupied by bee swarms. Soon after the first rains in spring the openings close and the bees are literally squeezed out. Thousands of swarms stray and try to settle anywhere, under the roof of buildings and sheds, in cars (beneath the motor cover) and in boots. A visitor left his hat on a porch; after a coffee break he found it filled with a bee swarm (De Kok 1976).

12.2.3 Nesting Sites

Lack of appropriate nesting sites, (not predation or lack of food) seems to be the limiting factor for bee populations in most of the tropical regions, except for the arid zones. This explains why, in spite of permanent intensive colony destruction, no reports of bee population decline are found. In any African country, east or west, the simplest way to start an apiary is to display empty boxes, as was mentioned before; in the season of reproduction they will soon attract swarms. "In this period the buzz of swarms is permanently to be heard" (in Zambia; Brown 1978). The lack of nesting sites of appropriate dimensions has two consequences: feral colonies in tropical Africa are generally small, occupying not more than five to eight combs. A new swarming cycle starts relatively soon because of overcrowded nests. In consequence, no professional queen rearing ever developed in subsaharan Africa. In S Africa even commercial beekeepers with 2000 colonies do not bother to replace colonies, in spite of heavy losses due to death of the queen, absconding, enemies, and theft (Fletcher 1978). Displaying empty boxes solves the problem easily.

12.2.4 Reproduction

For races with high swarming frequency, a large number of swarm cells would be assumed. In fact, the number is not greater in subsaharan, than in European races (5-15; Fletcher 1978a), while some Mediterranean races produce several hundred. Surplus virgins seem to be eliminated during the swarming process; N'Daye (1974) once found 80 dead queens underneath a big swarm (formed by uniting several during flight). Repeated swarming of the same colony (4-5 swarms) results in a reduced size; finally there are only about 2000-3000 bees. Yet even these minute swarms quickly develop to a full colony, as may the mother colony. "Main concern of the colonies (in Senegal) seems to be their multiplication, not the storing of reserves" (Dubois and Collart 1950).

12.2.5 Colony Growth

The surprising speed at which a colony may develop from a small swarm under favorable conditions is explained by the biological characteristics of tropical bees. Size, an important factor in endothermic temperature control (p. 166) is of less significance in hot climates. *Mellifera* bees of the tropics are of small size and, in consequence, the same is true also for the brood cells. While in European bees 800 cells are counted per dm^2, it is over 1000 in subsaharan races (Smith 1961; Chandler 1976), a plus of 25%. The spacing of combs is 29-32 cm instead of 35 mm (+10%). Adding the shorter development period of 18 1/2 days (Fletcher 1978, Moritz and Hänel 1984), African bees raise 50% more brood in identical volume and period than do European bees. African queens and nurse bees are evidently capable of coping with the ample supply of brood cells, since the daily egg rate of a queen during a nectar flow calculated from the brood present may be well over 2500 (Fletcher 1978a).

12.2.6 Seasonal Migration

In regions with important differences in altitude or humidity, migrations of colonies to locations with more favorable conditions were observed, as in *A. dorsata*, e. g., on river banks in Sudan (Rashad and El-Sarrag 1980), in the Rift valley of Kenya (Nightingale 1976), or on the slopes of the Kilimanjaro (Smith 1961). Colonies seem to be attracted from farther away by larger areas of flowering crop plants (Fletcher 1978). Several authors report on swarming and absconding during periods of dearth and the effect of water supply on the latter (e. g., Chandler 1976). Again, this mobility is peculiar for tropical races. In European bees, so-called "hunger swarms" are known (Zander and Weiß 1964), but generally a colony of the temperate zone will rather starve than leave the hive.

12.2.7 Field Activity

The flight of tropical bees is quick, seemingly erratic, and on a nonpredictable zig-zag course (Fletcher 1978; Kigatiira 1979), more resembling the flight of *A. cerana* from India than the steady and rather clumsy flight of *A. m. carnica* or *ligustica*. The selective advantage of this flight pattern was unintentionally demonstrated by N. Koeniger (p. 139).

Bees of East and South Africa use the same "dance language" for communication as do other races, but waggle runs start at a closer distance with rapidly decreasing frequency (Fig. 4.9; Smith 1958a; Fletcher 1978). This could indicate a shorter flight range, but the first author watched bees of his observation hive at a nectar source in 2000 m distance, and Fletcher (1978) was able to train bees to a feeding place at 1800 m.

Nights are cool in the interior and during the daytime strong radiation and high temperatures result in very low atmospheric humidity. This reduces availability of nectar to the hours from dawn to about 10.00 h and to the late afternoon. Surprisingly, African bees were observed to fly at lower ambient temperatures than did imported Italian bees. Repeatedly, full flight activity was observed at very low light intensities in the night during a nectar flow if the temperature was favorable (Fletcher 1978a).

12.3 Africanized Bees

In 1957 a limited number of queens from Transvaal and Tansania were imported to Brazil (Ribeirao Preto), in order to improve honey production by incorporating genes from adapted tropical strains (Michener 1975). By 1980 - in 23 years - descendants of these few queens had spread to the North of Venezuela, across 34° latitude and to N Argentina across about 14°. Feral colonies crossed the huge rain forests of the Amazonas, allegedly unfavorable to African bees (p. 217) and proceeded to the North with a speed of 100–300 km annually ($\bar{x} = 200$), completely wiping out earlier imported European strains.

Frequently these bees are referred to as "*A. m. adansonii*, "*A. m. scutellata*" or "African bees" (Kerr and Buono 1970; Nunamaker et al. 1984; Taylor 1985). Although these bees show the characteristics of tropical African bees (high swarming rate, migration, rapid colony build-up; Winston et al. 1983), it is not justified to homologize this bee with any known taxon, not even with the general term of origin as "African". The Africanized bee has lived for more than 50 generations in S America up to 1987, it was crossed with European stock and back-crossed and permanently selected by feral conditions in a tropical environment (Fletcher 1978a). The resulting bee can be distinguished morphometrically from *scutellata* of Transvaal (Buco et al. 1987). Therefore, the name "Africanized" seems most appropriate for a description.

A huge bibliography on the Africanized bee is now being summarized in a monography (Fletcher ed., in press). Therefore it seems sufficient to restrict the discussion of this difficult problem to a few of the fascinating biological aspects.

1. Two branches of evolution within the species *Apis mellifera*, the tropical and the temperate zone races, were abruptly confronted with each other in the biotope of one of them, namely the tropics. It became apparent in a brutal way that European strains do not adapt themselves totally to tropical conditions, in spite of living there for centuries. Only Africanized, but not European, bees were able to build up a feral population in S America. Characteristics important for survival in the temperate zone, such as low swarming frequency, immobility, being particular in choosing nesting sites, and avoidance of interactions among colonies, proved to be lethal in the confrontation with this specific competition. Interestingly, importation of European strains to the tropics of Asia involves problems with parasites and predators, but not disadvantageous interactions with congeneric species.

2. The outcome of the competition between the two *mellifera* branches was less a process of hybridization then rapid and complete elimination of the European bees. Hitherto unknown characteristics of African bees were brought to light: an aggressive reproductive behavior by invasion of other colonies, and parasitism through Africanized drones (Rinderer et al. 1987). This behavior has not yet been described for African bees in their home continent (Fletcher 1978a).

3. It is not yet known whether "temperate zone genes" were incorporated in the Africanized genom. It has to be stressed, however, that tropical and temperate zone characteristics are mutually exclusive. Present knowledge on the distribution of the Africanized bee in Argentina provides no evidence that this bee can survive in a temperate zone.

4. The African bees evidently found an "empty" ecological niche in S America, without heavy competition and with little pressure by predators. Therefore it was able to display a full potential for expansion. Colonies in Africa were observed to migrate "at least" 20 miles (30 km) to a rich nectar source. For the whole population to accomplish an average annual distance of 200 km, a much farther migration distance is necessary, even if three swarming cycles per year are assumed.

12.4 Taxonomy

Considering the very impressive common behavioral characteristics of tropical bees and, moreover, the rather uniform yellow pigmentation of the majority of races, it is not surprising that in the premorphometric period all bees of the subsaharan region were regarded as one taxon (Kerr and de Portugal Araujo 1958; Rothenbuhler et al. 1968). Superficially, the behavior of subsaharan bees seems everywhere to be the same, fashioned primarily by tropical *Apis* characters as described above, but individual differences are observed in any character, e. g., the tendency for absconding and swarming, defense behavior, etc. (Chandler 1976; Fletcher 1978a,b; Kitagiira 1979). Observations differing from the habitual pattern are easily overlooked, e. g., the description of beekeeping without protective clothing in Botswana by Clauss (1983) in a hot region, which should provoke fierce defense reactions according to one hypothesis (Fletcher 1978a). However, it is evident that correlations between climate and behavior are frequently observed.

Arguing from the modest amount of data available, it seems that the warmer the climate (constant high temperatures all year round), the more restless and mobile the bees become. Perhaps it is not by chance that the only countries with successful attempts at beekeeping with movable frames so far are in higher altitudes and latitudes (Ethiopia, Kenya, Tansania, South Africa). In Guiné-Bissau, Svensson (1984) concludes: "It is not worthwhile trying to work with movable frame hives".

Honeybee races of the temperate zone have their behavioral individuality, but beneath the homogenous surface imposed by the cool climate. The same could be true for bees of different climatic regions of the tropics. Only little can be said at present about such differences and it might be difficult to collect the needed data because of the irritability of the bees. Even in Transvaal, working with an observation hive is not without problems (Fletcher 1978a).

12.4.1 Morphometric Overview

The analysis of geographic populations of honeybees always started with a morphological description. The first morphometric study of honeybees of all of Africa, as presented in this section, does not aim to achieve a complete analysis of the variability found in this huge continent and to demonstrate borderlines of geographic types. Solid ground was only slowly provided by gradual accumulation of data, as

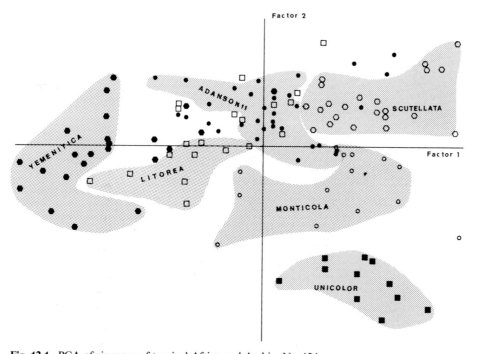

Fig. 12.1 PCA of six races of tropical Africa and Arabia. N = 124

shown by a series of publications with an increasing amount of information (Ruttner 1975, 1981; Ruttner and Kauhausen 1985). Although samples of all but a few African countries have been obtained by now (the missing countries are Angola, Liberia, Libya, and Sierra Leone), to date not more than a general outlook and a first attempt at classification can be presented. In the following paragraphs the geographic races established by multivariate analysis are briefly described and their discrimination discussed, using the graph of Fig. 12.1. As shown by Ruttner (1975, 1985, 1986b) and Fletcher (1978a), the isolating factor for races of honey-bees in tropical Africa is mainly climate. Along the east coast and in the mountains and highlands of East and S Africa, a textbook ecological isolation of races is found: four eco-races, each attributable to a certain climate (arid, coast, highland and mountain) can be clearly distinguished phenetically (Figs. 4.12, 4.13, 4.14). More complex, and today seemingly even contradictory, is the situation in all of W Africa. Although a phenetic N-S cline was established along the African west coast (p.53), no morphometric differentiation has yet been found, in spite of the huge geographic distance and important differences in humidity and altitude. More data and refined analyses are needed for better understanding. The same is true for the unexpected border between *A. m. yemenitica* and *adansonii* without substantial ecological changes in the center of the continent (Chad/Niger) (see Fig. 12.2).

Morphometrically, eight groups can be distinguished as listed below. Selected characteristics are shown in Table 12.2. All total analyses, in different combinations of factors 1, 2, and 3, result in basically the same arrangement located in the

Fig. 12.2 Map of Africa with climate zones. *Numbers* mean annual temperature. *Dense horizontal shading* subtropical winter rainfall region; *white* desert or highly arid; *oblique shading* tropical summer rainfall region; *horizontal shading* humid equatorial region; *vertical shading* equatorial climate region with two dry seasons. After Fletcher (1978a) and Walter (1958)

Table 12.2 Discriminant characters of eight subsaharan races of *Apis mellifera* (means and standard deviations of n samples, measurements in mm)

Character	*lamarckii* n=16	*yemenitica* n=30	*litorea* n=11	*scutellata* n=19	*adansonii* n=23	*monticola* n=9	*capensis* n=10	*unicolor* n=8
Length of hair	0.23	0.195	0.23	0.22	0.24	0.26	0.21	0.26
s.d.	0.04	0.02	0.02	0.04	0.02	0.04	0.03	0.04
Proboscis	5.81	5.48	5.81	5.86	5.68	6.06	5.80	5.67
s.d.	0.23	0.12	0.20	0.17	0.18	0.13	0.12	0.13
Fore wing l	8.38	8.13	8.40	8.66	8.45	8.85	8.95	8.77
s.d.	0.25	0.19	0.10	0.16	0.16	0.23	0.37	0.32
Hind leg	7.47	7.12	7.26	7.58	7.49	7.68	7.82	7.46
s.d.	0.27	0.22	0.12	0.20	0.15	0.22	0.37	0.14
Terg.3+4	4.24	3.94	3.92	4.17	4.02	4.17	4.25	4.03
s.d.	0.17	0.14	0.10	0.14	0.10	0.17	0.21	0.12
Stern 6 Ind	87.39	82.86	85.08	85.05	84.47	86.05	86.07	85.84
s.d.	4.80	3.38	2.99	4.05	4.44	5.80	3.95	2.35
Cub Ind	2.37	2.20	2.25	2.52	2.39	2.35	2.33	2.79
s.d.	0.37	0.40	0.41	0.46	0.41	0.41	0.34	0.42
Angle of wing venation I16	96.76	91.09	91.31	92.40	94.95	86.44	92.66	92.46
s.d.	3.60	4.89	4.33	4.19	3.82	8.37	3.81	3.24
Color terg 4	6.37	7.49	7.30	7.31	7.44	3.50	4.68	3.73
s.d.	2.32	1.27	0.80	1.32	1.26	1.92	0.59	1.11
Color scutell	5.58	6.26	6.77	5.61	6.13	0.97	1.93	0.35
s.d.	2.17	1.02	0.96	1.46	1.40	1.81	2.02	0.73

A-branch of the "Y" (Figs. 10.5, 10.6, 10.7). All races of tropical Africa, including *capensis* and *lamarckii*, are found in one closed group of clusters (with various levels of overlapping, depending on the kind of analysis) with the clusters *syriaca* and *sahariensis* next to it (Fig. 12.3).

A special problem is the classification of *A.m.lamarckii*. This race does not correspond to the behavioral pattern of tropical bees as described above: no absconding and migration, an unusually high number of swarm cells and polygyny during the time of reproduction, as found in other races of the eastern Mediterranean (p. 188). Yet, morphometrically, *lamarckii* undisputedly belongs to the group "tropical Africa", even to a section within this group with very small bees between *litorea* and *adansonii* (Fig. 10.5, 10.7, 12.5). Since in our analyses morphometrical data were used as principal criteria, we decided to include *lamarckii* with the tropical group, although it belongs to the Mediterranean geographically and behaviorally.

12.4.2 Statistical Characteristics of Honeybees of Tropical Africa

Variables extracted by stepwise discriminant analysis

No. of char.	Character	No. of char.	Character
1	Length of hair (t5)	14	Distance of wax plates
5	Femur	17	Fore wing, length
6	Tibia	19	Cubital vein a
7	Basitarsus, length	25	Wing ven. angle G18
10	Tergite 3 (long)	31	Wing ven. angle O26
11	Sternite 3 (long)	35	Color of scutellum

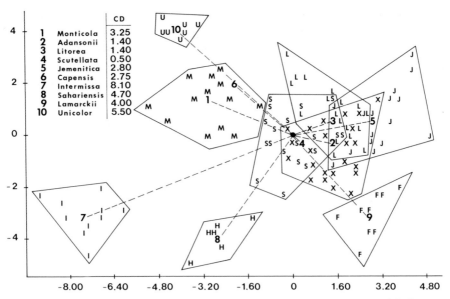

Fig. 12.3 DA of Africa. N = 170; *CD* canonical distances from common centroid of groups. *6 (capensis)* not shown

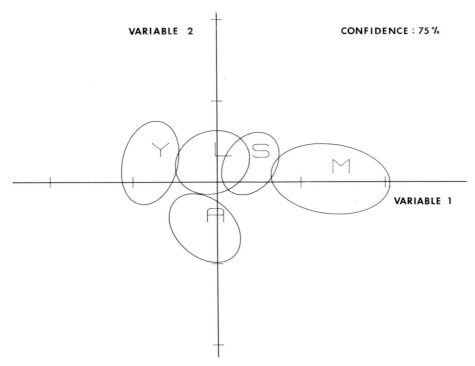

Fig. 12.4 Ellipses of confidence (75%) of DA from samples of five tropical African races. N = 124. A: *adansonii*; L: *litorea*; M: *monticola*; S: *scutellata*; Y: *yemenitica*

208

Statistical characteristics of races

Race	Canonical variable	
	Factor 1	Factor 2
lamarckii	2.77	−3.15
yemenitica	2.88	0.60
litorea	1.46	0.69
scutellata	0.35	0.17
adansonii	1.38	0.43
monticola	−3.01	1.30
capensis	−2.01	2.04
unicolor	−4.12	4.11
sahariensis	−2.71	−3.90
intermissa	−7.45	−3.05

12.5 Characteristics amd Natural History of Local Races

12.5.1 *Apis mellifera lamarckii* Cockerell (1906):166 The Egyptian Bee

This bee was first named by Latreille (1804) as *Apis fasciata*, but re-named by Cockerell because of "nomen preoccupatum". From drawings dated 2600 B.C, it is known that this was the first bee managed by men (Fraser 1951), using a technique which is still practised in Egypt now (Buttel-Reepen 1921). It has a long history in recent apiculture, since colonies of this bee were shipped to Germany, England, and N America as early as 1865–1867 (Buttel-Reepen 1906). In Berlin an "acclimatization association" was founded in 1864 with the special aim of importing bees from Egypt. The reason for this zeal in the apicultural world was the conspicuous color pattern of this bee: shining white, "silvery" hair on the thorax and tomenta, bright copper-yellow bands with shining black margins on the abdomen. This race is as beautiful in its color and even more than the Italians in their best beauty" Berlepsch 1873; Abushady 1919), it has "irresistible charm" (Br. Adam 1983). Various good reports on their behavior are available because of this early popularity.

A.m. lamarckii is not adapted to winter in the temperate zone. No true winter cluster is formed and as soon as it gets chilly, numb bees are found outside and inside the entrance (Buttel-Reepen 1921). Moreover, the flight activity is not blocked by cold on bright winter days.

In speculations on the evolution of honeybee races, A.m. lamarckii plays an important role. It is regarded as a "primary race" from which all yellow races of Africa, the Orient and also A.m. ligustica are derived (Buttel-Reepen 1921; Alpatov 1935a; Br. Adam 1983). These are ideas again based on similarity in color.

A.m. lamarckii is restricted to the narrow Egyptian Nile valley. No bees were found in Wadi Halfa (F. Schremmer, pers. commun.) and the zone without bees to the south (Sudan) is supposedly more than 500 km. The samples analyzed in this study were collected around Cairo, in Fayum and in Assiut (400 km south of Cairo). As already stated by Kaschef (1959), Wafa et al. (1965), and Br. Adam (1983), no regional variability is found within this race.

The Egyptian bee belongs to the smallest races observed and it is classified statistically among the subsaharan and not among North African races (Fig. 12.5) The visual impression of a "very little bee" is emphasized by an extreme slenderness, again a characteristic of a subsaharian race. The index of slenderness (St6) is higher than in any other race (Table 12.2). Hair length, proboscis, and length of forewing are similar to *litorea*, but tomenta are very broad (index 2.5) and the length of abdomen (T_{3+4}) is larger than in other races of tropical Africa and larger even than in *syriaca*.

A. m. lamarckii, therefore, can be described as a small, very slender, short-tongued, short-winged and short-legged bee. The drones (n = 4 samples) are smaller than in any other race examined, thus differing from the relative position of the workers (Fig. 10.8). In sample no. 797, the smallest of our collection, medium fore wing length is 10.46 mm, width 3.31 mm, length of hind leg 8.32 mm. Tergite 2 shows a narrow yellow stripe, but tergites 3 and 4 have only yellow spots.

Lamarckii appears as a compact cluster in any multivariate analysis, located between *yemenitica* (populations from Tschad and Sudan) and *syriaca* (Fig. 12.5). Phenetically, it takes a position between tropical African and oriental races.

In behavior, *lamarckii* is remarkable for its lack of propolizing, quite in contrast to its neighbors *syriaca* and *intermissa* (Buttel-Reepen 1921). Brood activity is high, continuing throughout the whole year, including maintenance of drones. Swarming tendency is not especially strong as can be concluded from the rare occurrence of afterswarms. Evidently, a less strong selection pressure for a high reproduction rate exists in the Nile valley than in subsaharan Africa and in parts

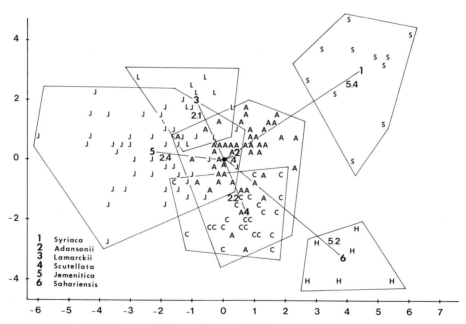

Fig. 12.5 Canonical distances of the two closest neighbors to the centroid of the compound cluster tropical Africa. *A. m. syriaca* (5.4) and *A. m. sahariensis* (4.2) are almost equidistant

of N Africa with its long dry spells. In contrast, in Egypt strong summer colonies may have a better survival chance against predation by hornets. A large number of queen cells are constructed (50–260 cells; Rotter 1920; El-Banby 1963a). As long as none of the young queens are mated, many virgins live together in the colony and in the swarm as they do in other races of the region (*syriaca, intermissa, sicula*). The author once collected 36 living queens out of a *lamarckii* swarm. Twelve queens were put together in a small cage. All of them were alive 24 h later. Worker bees start oviposition already a few days after dequeening (Buttel- Reepen 1921).

Time of development is somewhat shorter than in European races: 19.4 days for workers and 15.4 days for queens (El-Banby 1963b). The short wings probably are the reason for a higher wing beat frequency, since observers unanimously note a "distinctive sharp flight song" (Abushady 1919). The bees display an energetic defense behavior if disturbed. The question is open as to orientation ability, taking into account the stacks of several hundred colonies. It might be as well developed as in carniolans, but this has not yet been proved by investigation. Waggle dance starts already at a distance of 15 m and decreases quickly in frequency of waggle runs similar to bees of East Africa (Boch 1957; Smith 1958a), indicating a short flight range.

During the summer, the colonies suffer heavily from attacks of *Vespa orientalis*, and no special defense mechanism seems to exist. However, a very interesting observation was reported by Br. Adam (1983): in one traditional apiary the cylindrical hives were left open in front, thus completely exposing the combs which were covered with a compact cluster of bees. Any hornet venturing near was overwhelmed by flight bees starting from and alighting directly on the cluster without the risk of being killed. This reminds one of the defense of *A. cerana* against *V. mandarinia* (p. 136).

Along the Nile valley a subsaharan type of bee extends to the Mediterranean. Environmental conditions, however, are completely different. The Nile valley is irrigated and abounds with agricultural crops providing an adventitious flora which is completely different from the original one as shown by pollen analysis of a honey 3500 years old (Zander 1941). No special adaptations are required in this environment, and therefore, Egypt provides one of the few examples of a tropical local bee being easily replaced by temperate zone races. Surprisingly, no hybrids are found in traditional apiaries, although this race is known to hybridize freely with other races in Europe (Br. Adam 1983). A study of mating behavior could give the answer.

A note has to be added on the indigenous bee of Libya. "This bee has nothing to do with *intermissa*" (El-Banby 1977). It is reported to be similar in the color of the abdomen and the white hair of the thorax to the Egyptian, but larger and more docile. The biometrical data are not sufficient to clarify the possible relation with *A. m. sahariensis*.

12.5.2 *Apis mellifera yemenitica* Ruttner (1975):20

This most interesting race, one of the major subspecies of *A. mellifera*, extreme in morphology as well as ecology, was not recognized as a taxonomic unit of its own until 1975. Alpatov (1935a) mentioned a sample of bees from Yemen "which resembled bees from Egypt". Guiglia (1964, 1968) recorded "*Apis adansonii* Latreille" from Yemen, but neither of the two authors provided morphometric data. When H. Peters (Heidelberg) encountered honey as part of the food resources during an ethnological study in north Yemen in 1970, he believed the collected bees to be *A. cerana indica* because of their small size. A morphometric analysis revealed that it was a small *mellifera* bee of the subsaharan type (Ruttner 1975). The same type of bee was found also in Oman (Dutton et al. 1981), in Sudan (Ruttner 1975; Rashad and El-Sarrag 1980, 1984), Chad (Gadbin et al. 1979), Somalia, and Saudi Arabia. Therefore, what first seemed to be nothing but a local peculiarity (comparable to *A. m. major* of the Rif mountains) emerged as one of the "big" *mellifera* races characteristic for the hot arid zones of eastern Africa and Arabia. It covers a territory extending across 4500 km from East to West, comparing well with the territory of *A. m. mellifera*. *A. m. yemenitica* is exactly its opposite in almost every respect (hair length, size, proportions, color, and ecological adaptions). While *mellifera* reaches as far as the very limits of the genus *Apis* towards the polar circle, *yemenitica* takes the same place towards the hot deserts.

Fletcher (1978a) made an attempt to establish a correlation between climatic data and the distribution of honeybee races in Africa at a period when nothing was known about the distribution of the xerothermic *yemenitica*. He found many "white areas" on the map and took a guess, taking into account the elevated annual temperature "... either *A. m. nubica* or an unknown race may inhabit the Horn of Africa". Since *nubica* is now taken as a synonym for *yemenitica*, Fletcher's prediction turned out to be absolutely correct. On a map showing the mean annual temperatures, *yemenitica* is placed in the zone with the highest temperatures, ranging between 27–31°C (Fig. 12.2). Rainfall is very low (50–300 mm) and, what is more important, irregular. Plants and animals living in this zone have to be able to survive a whole year or even more without any rain.

Morphometrically, *yemenitica* is placed in the lowest sector of the total variability of size and hair length in *A. mellifera* (Figs. 10.7, 12.1), completely overlapping with *A. cerana* in this respect. Populations from the regions with the highest temperatures and lowest precipitation have the smallest bees (Saudi Arabia, Sudan). By PCA (factor 1 and 2) several subclusters are discriminated. However, if compared to neighboring races, they all belong to the same unit.

The behavior of *yemenitica* is little known, since only few attempts have been made to hive this bee in boxes with movable frames. The information published by Dutton et al. (1980) and Field (1980) on the bees of Oman and Yemen and by Rashad and El-Sarrag and by Wille (1979) on the bees of Sudan are summarized in the following paragraphs. As far as defensive behavior is concerned, the bees of N Oman and Sallalah as well as those of N Yemen are reported to be surprisingly docile. The bees of Sudan were found to react more quickly and strongly compared to *carnica*, but Wille experienced hardly any fierce attacks when investigating wild colonies in Sudan.

Reproductive swarming may be frequent in a good season (up to 10–12 swarms per colony in Oman). A rapid build-up of partial populations and re-colonization might be essential after a longer period of drought when in parts of the area bees could have completely perished. The number of swarm cells is reported usually to be between 10 and 15, as in tropical or European races, and thus differing from the E Mediterranean group. Absconding was not reported in Oman and Yemen, but during the dry season in Sudan. There the interesting phenomenon of "migratory swarms" was observed (Rashad and El-Sarrag 1980). The mother colony remained year round at the original site, in reach of a permanent water supply, while swarms migrated during the dry season. They do not necessarily build combs; if so, only honey but no brood is found in the cells. The bees of Chad are described as "migratory and aggressive", thus approaching the behavior of other subsaharan races (Gadbin 1976).

Local Populations

1. Saudi Arabia. The six samples from this country are from the SW (Sabya), the central (Riyadh) and the eastern part (Al-Hassa) of the country. They show high variability, but all are extremely small, slender, short-haired, and yellow. Sample no. 1157 represents the smallest *mellifera* bees measured: proboscis 5.02 mm, fore wing length 7.66 mm, hind leg 6.78 mm, color of T_2 9.0. This population of a presumably small number of colonies could be a relic of a more humid period.

2. Yemen and Oman. The south coast of Arabia is influenced by the monsoon; the mountains receive an average of about 300 mm (up to 600 mm in higher altitudes) of rain in the summer and a thornbush vegetation can exist thanks to the high atmospheric humidity at this time. In antiquity, the kingdom of Saba ("Arabia felix") provided the whole Roman empire with incense and other aromatic spices, and the writer Strabo also mentioned honey among the products of the country (Mandel 1980). At present about 40,000 hived colonies are counted in North Yemen (Field 1980). Beekeeping in Oman was described by Dutton and Free (1979) and Dutton et al. (1981), in Yemen by Field (1980). Colonies are kept in horizontal log hives (in Oman in hollowed-out trunks of date palms), in wicker-type basket hives or in wooden boxes. In Oman feral colonies are restricted to the mountains N of Salalah and the Jebal Akdhar (alt. 2980 m). Main honey plants are *Ziziphus spina-christi* and *Acacia tortilis*. The assumption that bees had been imported to N Oman by man only 200 years ago is questionable since the detection of autochtonous honeybees in Saudi Arabia. Disjunct areas can easily be explained by changes in climate.

The worker samples from Oman and Yemen (n = 21) form a solid cluster between "Saudi" and "Chad", although the Omani bees are slightly bigger than those of Yemen. The drones from Oman and Yemen (n = 14 samples), the only ones of

A.m.yemenitica available, belong to the smallest of the species, together with *adansonii* and *lamarckii*. Color of tergite 2 is highly variable, ranging between all dark (1.0) and largely yellow (6.2).

3. *Sudan*. The canonical variables of the bees of Sudan (five samples) are well separated from the clusters of neighboring races (*lamarckii, litorea, scutellata, monticola*), justifying the establishment of a separate race, "*nubica*" (Ruttner 1975).
Later it was found that they are completely scattered among samples of Yemen,
Somalia, and Chad. Therefore, the name "*nubica*" has to be cancelled – one of
many mistakes due to lack of knowledge – and this taxon is incorporated into
the subspecies *yemenitica* as a Sudanese population.

Rashad and El-Sarrag (1980, 1984) published thorough morphometric surveys of
the bees of Sudan. They found a high degree of regional variability (consistent
with our data) and finally established two subspecies: *A.m.sudanensis* in the
major part of the country, which corresponds very closely to *A.m.yemenitica*, as
described above (small size of all parts of the body, slender, very yellow);
A.m.nubica along the southern borders of Sudan, described as larger and darker
(this type might be very close to *A.m.scutellata*).

In general, the Sudanese bee has slightly larger body dimensions than the
yemenitica populations from Arabia but rather a more slender and brighter abdomen (Table 12.3). Feral colonies are widespread in the southern part of the country, nesting in trees, logs, rocks and termite mounds, but very little apiculture
seems to exist.

4. *Somalia*. Bees are found mostly along rivers and considerable honey yields can
be obtained (Sartorelli, pers. commun.). The workers ($n = 9$) are exactly the same
size as the bees of Sudan, with slightly longer proboscis and broader and darker
abdomen.

5. *Chad*. Gadbin et al. (1979) and Gadbin (1980) studied bees and honey plants of
the southern Chad. The bees were determined as belonging to *A.m.adansonii*,
but this was in a period where any small and yellow African bee was named

Table 12.3 Characteristics of five populations of *A.m.yemenitica* (Mean and standard deviation; in mm)

Population	n	Terg 3+4	Proboscis	F wing length	Hind leg	Stern6 I	Hair length	Cub I	Angle J 16	Color terg 4
Saudi	6	3.748	5.277	7.868	6.916	84.71	0.172	2.28	89.94	4.60
		0.153	0.210	0.224	0.259	2.78	0.021	0.25	2.90	0.99
Yemen,	30	3.937	5.481	8.135	7.120	82.86	0.195	2.20	91.09	4.52
Oman		0.137	0.132	0.192	0.219	3.39	0.020	0.40	4.16	1.27
Somalia	9	3.981	5.552	8.214	7.207	85.07	0.213	2.27	99.33	7.75
		0.121	0.120	0.179	0.203	2.63	0.017	0.36	8.03	1.03
Sudan	5	3.965	5.450	8.219	7.204	89.31	0.193	2.45	92.60	6.38
		0.180	0.187	0.214	0.245	4.50	0.033	0.42	3.49	1.15
Chad	8	3.914	5.356	8.136	7.175	88.15	0.211	2.39	95.90	5.36
		0.121	0.187	0.141	0.205	4.89	0.019	0.38	3.96	1.11

"*adansonii*" and no reference samples of neighboring countries were included in the analysis. Although still belonging to the Sahel zone, annual precipitation exceeds 500 mm in the south of the country and several north-bound rivers permanently carry water. Therefore, seasonal differences in nectar supply are less pronounced than in the other regions mentioned.

Samples from Chad, kindly provided by the same author and incorporated into our collection, clearly proved to be *yemenitica,* again solidly clustering as do other local populations of this race, with close relation to the samples of Sudan and Somalia. Differences in the morphometric values of these populations are minute (Table 12.3).

12.5.3 Apis mellifera litorea Smith (1961):259

This is another only recently established race of tropical Africa, occurring along the east coast of the continent from Kenya to Mozambique. It is easily recognized by its small size compared to the savannah bee, *A.m.scutellata,* although for decades subsaharan bees were assumed to belong to the same geographic race. In the dry climate of the Somalian coast *litorea* is replaced by the still smaller *yemenitica.* The coastal plain of Kenya and Tansania being rather narrow, extensive hybridization with *scutellata* from the highlands has to be assumed. The largest population of *litorea* probably exists in the plains of Mozambique.

Morphometrically, *litorea* takes an intermediate position in size and hair length between *yemenitica* (to the north) and *adansonii* (on the West coast). The immediate neighbor, *scutellata,* is much larger. A specific characteristic is the long proboscis relative to body size and leg length (Table 12.2). As to pigmentation, no significant difference exists with other subsaharan races. Mean worker cell size was measured to be 4.62 mm, spacing of brood combs 28–30 mm (Smith 1961).

The climate of the African east coast is warm and humid, inducing almost continuous flowering and brood rearing throughout the year, but in periods of dearth absconding is common as soon as the stored food is exhausted. The same behavior is triggered by attacks of the wasp *Palarus latifrons* (bee pirate), nesting in the sandy soil.

The sequence *litorea-scutellata-monticola* provides an excellent example of a graded, stepwise ecocline (Chap. 4)

12.5.4 Apis mellifera scutellata Lepeletier (1836):404 (syn. A.m.adansonii Latr. 1804)

The original *scutellata* sample of Lepeletier is labeled "South Africa: from Caffreria". Therefore, this name is best suited for the bee of the East and the South African thorn woodland and tall grass savannah ("miombo") found from Ethiopia to the Cape Province. Formerly classified together with *A.m.adansonii,* the Westafrican bee described from Senegal (see p.216), it clearly deserves the status of a separate subspecies, as shown by the morphometric data (Table 12.2, Figs. 10.5, 12.1,12.3,12.4). It is found in the highlands at altitudes between 500 m and 2400 m, between the range of *litorea* and *monticola* (Smith 1961). Samples belonging to the *scutellata* cluster were investigated from eight countries (Ethiopia, Kenya, Uganda, Tansania, Rwanda-Burundi, Malawi, Zimbabwe, and S Africa).

Ecologically, the *scutellata* area is characterized by two dry periods between relatively cool rainy seasons. Mean annual temperature between 16° and 23° (Fig. 12.2). This area is clearly distinguished from the much warmer *litorea* and the cool *monticola* area.

Within the group of tropical bees *scutellata* is relatively large, corresponding to the altitude of its biotope. When compared with *litorea*, all dimensions are about isometrically increased *except* for the slightly longer tongue (Table 12.2). More complex are the differences from *monticola*; the dimensions of the body (tergites, sternites) are almost identical in both races. However, proboscis, fore wing, and hind leg are considerably longer in *monticola*; these dimensions increase with altitude, a cline not corresponding to Allen's rule.

The drones (nine samples from Kenya, Tansania, and Burundi) take a central position in a PCA, close to *monticola* (Fig. 10.8). They have surprisingly long fore wings and hind legs (compared to *adansonii* and to N African races), and are totally black.

Generally, *scutellata* is regarded as *the* African bee, since several countries of the region are well known and beekeeping is practised also with movable frames. Therefore, almost all the knowledge on physiology and behavior of subsaharan bees was obtained from this race. Reports have been published from Ethiopia, Kenya, Tansania, Zimbabwe, and S Africa. They concern mainly general characteristics of tropical bees and since detailed comparative data are lacking for the other races of tropical Africa, these observations were included in the general introduction to this section.

12.5.5 *Apis mellifera adansonii* Latreille (1804):172

Michael Adanson, a student of Reaumur in Paris, stayed in Senegal from 1749-1754, and among numerous plants and insects, collected also bees. His "Histoire naturelle du Sénéegal", Paris 1757, is one of the first scientific descriptions of a tropical country.

The history of this taxon with its vicissitudes is described in detail in Chap. 5. As a result of the confusion, bees of various countries, including the Africanized bees of S America, but only very few of the correct subspecies of West Africa, were described under the name "*adansonii*". It took many years to collect sufficient data to reduce the name to its original significance: a subspecies of West Africa only. Even now important questions remain unanswered, as will be outlined later.

Adansonii is the African race with probably the largest area of distribution. Samples of most of the countries between Senegal-Mali-Niger in the north to Zaire in the south were available for morphometric studies. No bees were analyzed from Angola, and little can be stated about the demarcation from *scutellata* in the center of the continent. At the borderline towards *yemenitica*, the situation is hard to understand: in the south of Chad typical *yemenitica* is found (p.211). In Niger, the western neighbor, however, a bee is found which matches in every detail the mean values of *adansonii*. The climate of these two countries of the Sahel zone shows no essential difference – at any rate it is less important than the diversity within the *adansonii* region. Further studies are needed to be able to explain this unexpected transition from one race to another.

Some specific characteristics of the subspecies are shown in Table 12.2. *Adansonii* is a small, yellow bee, but somewhat larger than *litorea*. Its cluster in PCA (factor 1 and 2) is found between *litorea* and *scutellata*; much overlapping is evident. Perhaps discrimination can be improved by including drones in the analysis. *Adansonii* drones (six samples from Senegal, Guinea and Ghana) are small, second in size only to *lamarckii*. They belong to the few races with frequent yellow stripes on tergite 2 and 3 (together with *sahariensis* and *lamarckii*), although sometimes black drones are observed also (in Angola, de Portugal Aranjo 1956). Drones of East Africa (*monticola* and *scutellata*), however, were found to be all black. Unfortunately, no data are available yet on *litorea* drones. At the level of 75% confidence the clusters *adansonii* and *scutellata* are well separated (Fig. 12.4).

The African west coast is more humid than the east coast, with only one dry period instead of two per year. Rain forests occur in wide river valleys and basins (Niger, Congo) and along the coast. Bees are abundant everywhere, but foremost in the savannah. According to the corresponding personal information of two experts, C. H. Fry (Aberdeen) and R. Darchen (Les Eyzies), honeybees are mainly observed in the open country or in gallery forests along rivers. The question has still to be answered whether in large areas of homogenous rain forest, e.g., in the Congo basin, a considerable bee population exists. The most important nectariferous plants of Nigeria and Chad were assessed by palynological studies of honey (Agwu and Akanbi 1985; Gadbin 1980).

The phenetic similarity of *adansonii* and *litorea*, and the similarity of occurrence on both coasts of tropical Africa suggest ecological analogies. However, this conception does not take into consideration that climatic contrasts exist in West Africa: only the coastal zone is humid, while in the interior, north and south of the tropical rain forest, a dry savannah is found (Fig. 12.2). *Adansonii*, however, is not limited to the coastal zone as is *litorea* in the east, but it occurs as far inland as Upper Volta and Niger, colonizing semi-arid zones without phenetic changes. In the case of *adansonii*, therefore, no evident correlation between phenotype and environment, as found in other African races, can be demonstrated (at least at the present state of knowledge).

In contrast to South and East Africa, no scientific investigations exist about the behavior of the bees of West Africa. However, a number of detailed reports on apicultural projects transmit various useful informations (N'Daye 1974, Senegal; Dubois and Collart 1950, Zaire; Lepissier 1968, CAR; Brown 1978, Zambia; Svensson 1984, Guiné-Bissau; Nunes and Tordo 1966, Angola). All of them stress the overall presence and the abundance of bees (although in varying intensity) in the countries of western Africa. The overall behavior corresponds to the description given in the introduction to this section on bees of tropical Africa (p. 199). As far as the defense behavior is concerned, a certain differentiation can be observed. First, differences in irritability are reported between different regions of the same country; second, the method of management and a process of habituation are of importance. In Lusaka, bee colonies can be kept close to the house, but the day after honey removal the bees are highly irritated and prone to sting persons passing by (Brown 1978). Villages exist in Guinne-Bissau where bees are kept in cavities in house walls and bees fly to and fro over the head of playing children; combs of honey are taken carefully from behind while smoking the bees; however, working with Langtroth hives seems not advisable because of the irritation of bees during manipulation of the combs (Svensson 1984).

The whole scale of exploitation of bee colonies is found in West Africa, as in other parts of subsaharan Africa: pygmies of the Congo basin close the entrance of a colony with fresh Manioc leaves, killing all bees in (Dubois and Collart 1950). In Senegal, tubes made of elephant grass are placed in trees to attract swarms; to harvest the honey at night, the colony is shaken above a fire until all the bees are killed in the flames. However, in most countries a traditional apiculture is practised in some districts. In the Central African Republic 300,000 log or wicker hives were counted in 1968 (Lepissier 1968). Frequently the brood is destroyed when harvesting the honey, but not the bees. However, no report was found on successful apiculture with movable frames in this part of the continent.

12.5.6 Apis mellifera monticola Smith (1961):258

Monticola takes a special position among honeybee races: it is the first taxonomic unit demonstrating isolation entirely by ecological factors showing a unique disjunct area of distribution. Monticola is the bee of the rain forests of the East African mountains found in altitudes of 2000–3000 m. The mean annual temperature in the rain forest of the Kilimanjaro (alt. 2600 m) was determined to be 11.2°, whereas the mean annual temperature of a place 1500 m lower (Moshi, 1150 m) amounted to 20.8° (Walter and Breckle 1984). On many days of the year a belt of dense fog and clouds covers this zone, but on clear days strong radiation results in considerable temperature augmentation in plants (and certainly also in the body of flight bees. In spite of this cool climate, monticola cannot be compared with bees from the temperate zone, since longer periods without flight activity most likely do not occur.

Smith (1961) observed monticola on Kilimanjaro and Mt. Meru (Tansania) at altitudes of 2400–3000 m and he mentions having seen a few specimens of the same type from Ethiopia. We collected or received samples of the monticola type from the following locations (13 samples):

Location	Country	Altitude(m)
Kilimanjaro	Tansania	2400–3000
Mt. Meru	Tansania	2400–2600
Karemba	Burundi	1600
?	Ethiopia	"Highland"
Mt. Elgon	Kenya	2600
Nyambeni Range (Meru)	Kenya	2600
Aberdare	Kenya	3000
Mt. Kenya	Kenya	2600

The true area of distribution of *monticola* might be much larger than indicated by this list. Dubois and Collart (1950) report on "black bees" in the mountains of northern Zaire.

Holotypes of *A.m.monticola* collected by F.G.Smith in 1965 (four samples of the collection in Oberursel) can be described as a medium-sized bee, larger and with longer hair than all other races of tropical Africa (Table 12.2). One of the samples of the holotype group (no. 178) equals measurements of the temperate zone races: proboscis 6.55 mm, fore wing length 9.18 mm, hind leg 8.00 mm. Most of the bees show dark pigmentation of the abdomen, with small yellow spots, but a narrow yellow stripe on tergite 3 is frequent. One "African" characteristic, the very slender abdomen is common to all *monticola* colonies (index St6 85.0).

A difficulty in classification, overcome by the important morphometric distance to the neighboring *A.m.scutellata*, is the unusual variability of characters (see s.d. in Table 12.2). This distance is small or nonexistent in measurements of the body itself, but considerable in the appendices (tongue, hind leg, fore wing). The variability of the subspecies is found within a local population (e.g., Kilimanjaro), as well as in the whole geographic range, covering a distance of about 1500 km. This reflects the genetic structure of the subspecies:

1. *A.m.monticola* consists of numerous completely isolated populations (the large *scutellata* population round each *monticola* island provides a more efficient isolation than anything else). The present situation is probably the relic of a large, more or less coherent population during the last pluvial, since the snow line was 1000 m lower at this period with expansion of the rain forest (= the *monticola* biotope) across most of the highland savannah of E Africa (Schwarzbach 1974).
2. A perennial hybridization takes place with *scutellata* in a transitory zone of altitude. This zone is fluctuating, since *scutellata* colonies were observed to migrate up the Kilimanjaro slopes in the dry season and down again in the wet and cool season (Smith 1961). The "*monticola*" type has to be restituted again and again by selection via the ecological conditions. Some characteristics may be more sensible to this process than others; in Ethiopia a *monticola* colony with completely yellow but large bees was found (color T3 8.9, proboscis 6.12 mm, forewing 8.93 mm); on the other hand, one of Smith's samples from Kilimanjaro is completely dark, but very close to *scutellata* in size. Only a slight correlation seems to exist between size and color. No wonder that the *monticola* cluster is unusually large in a PCA (Fig. 10.5), and sometimes the classification of a sample may be somewhat arbitrary.

A rather sensible equilibrium is evidently maintained between the savannah and the mountain population. In 1984, 23 years after Smith, a sampling tour was made on Kilimanjaro and Mt.Meru together with G.and N.Koeniger. Again, a clear gradient was found for characters of size and color from 900 m (below Arusha) to 2800 m, but within a smaller range than before. It seems that now *scutellata* and hybrids penetrated further into the mountain region. The period of drought experienced at this time could have resulted in a temperature rise in the humid belt and also in a change of the bee population.

Monticola drones (two samples) are uniformly dark, with very small bright spots on tergites 2 and 3 and scutellum in about half of the specimens. In size they are second behind *unicolor*, thus reversing the established order in worker bees. Especially long are the hind legs of drones (9.78 mm) equalling larger Mediterranean races such as *ligustica* and *sicula*.

In contrast to the common image of the African bee, *monticola* is described as very gentle and even manageable without protective clothes (Smith 1961; Drescher 1975). It is reported to fly at lower temperatures than *scutellata* and to conserve stores during nectar shortage by drastically reducing brood rearing. It is surprising to find traditional hives in the virgin forest near its upper limit on Kilimanjaro and Mt. Meru, many hours walk from the nearest human settlement, indicating a good honey yield in this region. All of this seems very promising for apicultural use of *monticola*, but what was observed with *scutellata* in higher altitudes could be true also in the reverse situation, concluded from one experiment reported from Ethiopia (Gebreyesus 1976): *monticola* colonies transported to rich nectar sources in lower altitudes "soon start to dwindle and become lazy".

12.5.7 *Apis mellifera capensis* Escholtz (1821):97

In 1912 a hitherto unknown beekeeper, J.W. Onions, reported on a honeybee of the Cape Province (South Africa), which had to appear as pure phantasy to everybody experienced in the biology of honeybees:

The worker bees of these colonies, if deprived of their queen, would start to lay eggs within a short time and from these eggs hardly any drones would develop as expected, but mainly females (worker bees or queens). Onions was well aware of the fact that he was challenging the general belief in the exclusively arrhenotokuous parthenogenesis in honeybees and therefore he asked a South African entomologist, J.Jack, for scientific verification. Jack (1916) fully confirmed Onions' statements, as other researchers did later (Anderson 1963; Ruttner 1976c, 1977a, b). Thus the Cape bee established its permanent place in the literature as a singular biological curiosity.

In a queenless colony of *A. capensis*, laying workers occur within 3–4 days after disappearance of the queen, earlier than in any other race except *A. m. intermissa* (p.46). Frequently several eggs are found in one cell. With few exceptions all these eggs develop into worker bees appearing normal (Fig. 12.6), even if they were laid in drone cells. The scattered broodnest pattern is similar to that of laying workers in other races (Fig. 12.7). The worker larva may be raised to a queen, which may mate and restitute the colony into a queenright status again (Onions 1912; Ruttner 1977).

Astonishingly, a virgin queen behaves differently when she starts egg laying without mating. She produces part male and part female brood. In mated queens, however, male and female determined eggs are distributed in drone and worker cells as regularly as in other races. The laying worker is treated like a queen, is regularly fed, and the workers form a court around her (Fig. 12.8). The supposition that these workers are able to produce queen substance (9-oxo-decenoic acid) was confirmed by Ruttner et al. (1976) and by Hemling et al. (1979). The average amount of 9-ODA found in the head of a laying *capensis* worker is more than half of the quantity in a true queen. It increases with the age of the worker bee. As

Fig. 12.6 *Apis mellifera capensis,* worker bee

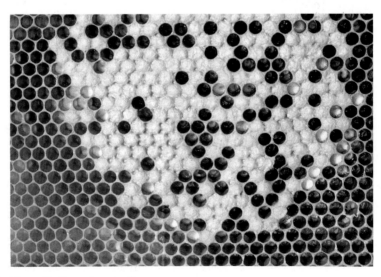

Fig. 12.7 Brood pattern of a "pseudo-queen"

Fig. 12.8 Laying *capensis* worker (white mark on the thorax) with court

soon as one worker starts with egg laying, ovary development is suppressed to a great extent in the other worker bees. Thus it is fully justified to speak of a "pseudo-queen".

The morphological basis of the phenomenon is better development of the ovaries (10–20 ovarioles in Cape workers vs. only 3–5 in other races; Fig. 12.9), a relatively large spermatheca as shown in Fig. 12.9 ($\emptyset$ 0.3–0.6 mm; Onions 1912; Anderson 1963; Ruttner 1977) and a bigger mandibular gland (Fig. 12.10). The volume of the mandibular gland of *capensis* is about double that of *carnica* (0.544 mm^3 vs. 0.275 mm^3). It increases slightly as soon as *capensis* workers start egg laying. The liquid inside the spermatheca seems to be of functional composition as concluded from the observation of mobile spermatozoa in this fluid after instrumental insemination (experience of the author; Woyke 1980). Under natural conditions the spermatheca of Cape bees was always found to be empty. In the external morphology *capensis* workers, however, do not show any queen-like or intermediate characteristics.

In experiments with several genetic markers it was demonstrated that segregation of genes occurs in this type of parthenogenesis (Kauhausen 1984). The cytology of the thelytokuous reproduction was studied by S. Verma and Ruttner (1983). The parthenogenetically produced workers proved to be diploid; meiosis started quite normally, with clear chiasmata, and four nuclei originate as the result. Diploidy is restored in late meiosis II by central fusion of the egg pronucleus with one of the descendants of polar body 2 (Fig. 12.11).

A young *capensis* worker kept together in a cage with 30 nurse bees of Euro-

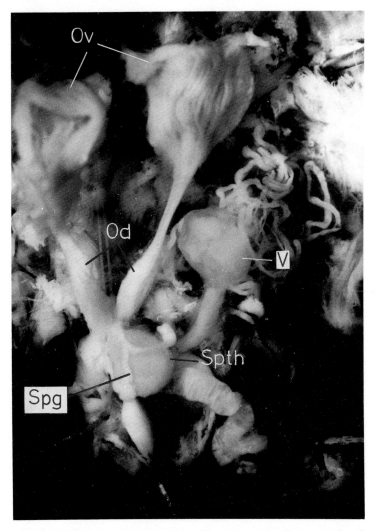

Fig. 12.9 Genital organs of a laying *capensis* worker. *Od* oviduct; *Ov* ovary; *Spg* spermathecal gland; *Spth* spermatheca; *V* venom sac

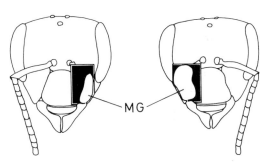

Fig. 12.10 Mandibular gland (*MG*) of *A.m.carnica* (*left*) and *A.m.capensis* (*right*)

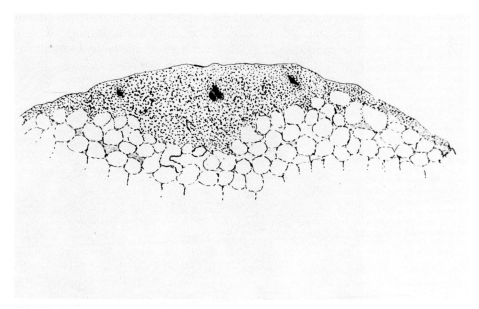

Fig. 12.11 Central fusion of two haploid nuclei during meiosis II in a worker-laid egg of *A.m.capensis.* (S.Verma and Ruttner 1983)

pean race will start to lay eggs very soon. Adult Cape bees can be raised from these eggs by appropriate techniques. This procedure may be repeated often and a long, uninterrupted series of parthenogenetic generations can thus be obtained, as was demonstrated by the experiments performed in the bee laboratory at Oberursel, W.Germany, and Utrecht, Netherlands. The time of development from egg to adult is remarkably shortened. The post-capping stage of this race was found to be only 9.6 days, 2.4 days shorter than of *carnica* brood used as controls (Moritz and Hänel 1984).

The Cape bee is an excellent tool for genetic studies of honeybees. While virtually all the characters studied in genetic and behavioral research are manifested in the worker caste only, the obligatory insertion of sexuals has been a heavy obstacle. With the thelytokuous workers the direct transmission of genes from one to the other worker individual is feasible. As thelytoky in the Cape bee is evidently based on one single recessive major gene (p.48), it is easily combined with other genes. This opens broad aspects, especially in population and behavioral genetics (Brandes and Moritz 1983; Brandes 1984; Moritz and Klepsch 1985).

Laying *capensis* worker bees are kept with young bees of European race in small nuclei in the open or in a flight room or in cages in an incubator. The minimum number of nurse bees for rearing brood to the adult stage is about 40-50. A few freshly hatched nurse bees are added every week. Laying workers are very long-lived (5 months and longer) even when kept in cages. However, they produce eggs only during a limited period of 4-56 days (mean 30 days). In rare cases favorable conditions (free flight, young attendant bees) may trigger a second oviposition period.

A. m. capensis was first described (and named) by Escholtz in 1821, and later by Alpatov (1938c) and Ruttner (1975, 1977a, b). Its external morphology places it among the subsaharan races (Fig. 12.3). This fact is easily shown by its small size and slender abdomen (Fig. 12.6). The dark body color (conspicuously different from its yellow neighbor *A. m. scutellata*) and larger size bring this race close to *A. m. monticula*. In general the scutellum is dark also. The dimensions of the abdomen and wings are slightly larger than those of the two other races of the African coast (*adansonii* and *litorea*), except for its short legs. *Scutellata* from higher altitudes is definitely larger in all dimensions. It shares the remarkable slenderness of the abdomen (St 6-index 84,7) with the other races of tropical Africa. Compared to its small body size, *A. capensis* is a long-tongued race (6.20 mm). According to Woyke (1976), another discriminant character is a higher bristle density on the surface of the fore wings. The scattered arrangement of our *capensis* samples indicates a considerable hybridization.

In spite of the major differences in internal morphology and biology *A. m. capensis* is nothing but a geographic variety of *A. mellifera*. This is shown by free hybridization with various other races (including those of Europe) and the fully vital and fertile offspring resulting from it. *Capensis* is an endemic race, restricted in its pure type to the Cape Peninsula and the immediate environment of Cape Town (Fig. 12.12). Transitory types are also found further North and East (Moritz and Kauhausen 1984). The true *capensis* population was estimated not to exceed 10,000 colonies (Ruttner 1977).

The population of unhybridized *capensis* gradually seems to be declining due to permanent importations to the region of *scutellata* stock. In 1983, pure *capensis* was found in Cape Town only (Moritz and Kauhausen 1984), in a more restricted region than given by Ruttner (1977). The limits indicated by Kerr and de Portugal Araujo in 1958 are by far more expanded (Fig. 12.12).

After loss of the queen, severe fights take place in the colony before one or several pseudo-queens are established ("rejecting behavior", Anderson 1963). Furthermore, Cape bees may intrude into a queenright neighboring colony and start

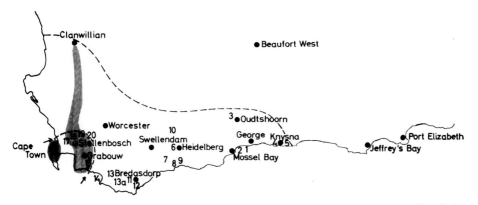

Fig. 12.12 Territory of *A. m. capensis* on the SW corner of the African continent. *Dotted line* limits as given by Kerr and de Portugal Araujo (1958); *line between arrows* limits according to Ruttner (1977); *hatched zone* winter rainfall region. (Ruttner 1977)

egg laying there. After a while the original queen will vanish and the colony is very difficult to re-queen. Thus thelytoky creates some problems in respect to the social organization of the colony.

How could this particular character evolve and persist in the history of this bee? An answer might be found in the unusually stormy climate of this region. If greater losses of young queens occur during mating flights and if these losses are replaced at least in part by new queens raised from pseudo-queens, thelytoky might be a selective advantage in spite of the problems it causes. Losses of colonies due to loss of the queen are very rare in the Cape region, as beekeepers report. Swarms may take off without a queen, and the bees will frequently raise queens afterwards from thelytokously produced eggs (Gough 1928). *A. m. capensis* is the only honeybee which colonized a region with temperate climate in the Southern hemisphere. Thus races of *A. mellifera* were able to adapt to cool climates south as well as north of the equator. The adaptation of *A. m. capensis* to a cooler, more humid climate was experienced by local beekeepers in Cape Town. They stated that *scutellata* bees – which are frequently exposed to low temperatures for a short time, but not to long periods with high humidity without flight – which were often imported to the Cape Peninsula disappeared within 1 or 2 years. Another adap- tation to a moderate climate compared to tropical bees is the diminished tendency to abscond.

Nevertheless, *A. m. capensis* is not fit to winter in Central Europe without applying sophisticated apicultural techniques. One reason for this (similar to bees of true tropical origin) is the lack of the "cold-avoidance behavior" (see p. 35). *Capensis* is relatively gentle and therefore is used as pollinator in South African orchards.

12.5.8 *Apis mellifera unicolor* Latreille (1804):1688

At what time a honeybee race received a scientific name in the history of bee taxonomy depends on chance. The local South African race of Madagascar was among the first to be described, probably because of its uniform deep black pigmentation, which is in sharp contrast with the colorful bees of the subsaharan part of the continent. Morphometric analysis confirms the early taxonomic classification of this island type, locating the cluster within the group of races from tropical Africa (Fig. 12.1).

Madagascar is situated between 12° and 25° S lat. Since the major part of the island consists of highlands between 1000 and 2000 m with average annual rainfall of about 1500 mm, the climatic conditions are comparable to the savannah of the African continent, except for less contrast. The analogous race of the continent, *scutellata*, is distinctly larger (tergites, hind legs, proboscis) except for its shorter fore wings. The specific differences between these two highland races, besides color, are those of allometry: compared to body size, *unicolor* has a short proboscis and very long fore wings (Table 12.2). The black pigmentation which includes the scutellum is the more conspicuous as the tomenta are very scarce and only indistinctly visible. The wings show a dark tinge, as already noticed by Maa (1953).

The climate is much warmer and more humid on the coast than in the highlands (Grünewald et al. 1983):

Toamasina (5 m), mean ann. temp. 24.2°, precip. 3530 mm
Antananarivo (1310 m), mean ann. temp. 17.7°, precip. 1361 mm

Two different ecotypes exist in these two zones, according to Douhet (1965): the coastal bees behave as do tropical bees, they have a high tendency to abscond (p. 201). The highland bees are described as being similar to European strains, they are stationary and very gentle, thus they can be kept within human settlements without any problems. Our collection of Madagascan bees consists of eight samples from the region of Antananarivo with characteristics as given in Table 12. Two samples are from the island of Mauritius; these bees are clearly *unicolor*, but distinctly smaller. They could be taken to be a black variety of *litorea* from the opposite coast of Mozambique were it not for the typical *unicolor* allometry: short proboscis (5.31 mm vs. 5.81 mm in *litorea*!) and long fore wings (8.50 mm vs. 8.40 mm). It is documented that the Mascarene Islands were colonized with bees by French settlers coming from Madagascar: Reunion in 1666 and Mauritius after 1721 (Bronard 1973). Both islands seem to have been void of bees prior to these dates. The period since then is too short to develop changes of morphological characteristics. Therefore it can be assumed that the bees of the islands are descendants of the coastal population of Madagascar described earlier. If this assumption can be verified by analyzing original samples from the Madagascan coastal area, a striking analogy with the African continent could be demonstrated: small bees with "tropical" characteristics on the coast and larger bees in the highlands.

Of all those investigated, *unicolor* drones (five samples from Atananarivo) are the darkest; tergites and scutellum have almost no bright spots. Surprisingly, in spite of the small size of worker bees, they are larger than in other African races, except for the relatively short hind legs. In an overall PC analysis of all *mellifera* races *unicolor* drones take a central position (factor 1, Fig. 10.8).

Only few reports exist on honeybees and beekeeping in Madagascar, with only brief remarks on biology. The behavior of *A.m. unicolor* has yet to be studied in more detail.

CHAPTER 13

Honeybees of the Western Mediterranean

13.1 Introduction: The Mediterranean Basin

The great variety of *mellifera* subspecies found around the Mediterranean fully justifies consideration of this basin as the main gene center of the species. Out of 24 subspecies described in this overview, 13 are based in the coastal areas of the Mediterranean. The races of the cold temperate zone originated and are still existing there and even a tropical race (*A. m. lamarckii)* was able to advance this far. The mild, humid climate and the rich flora of the region evidently favor the needs of honeybees (as well as the abundance of other Apoidea). The Mediterranean flora is well known for its uniformity. In contrast, the rich resources, minor varieties in climate and a fluctuating history during the quarternary (with very cool periods) promoted an ample radiation within *A. mellifera* (Chap. 3). Although most subspecies of the region are of medium size (Fig. 4.2), almost the whole range of size variation is found between *lamarckii* and *syriaca* on one end and *mellifera* and *carnica* on the other (Fig. 10.5). In a DA covering 9 Mediterranean races the variation extends from *mellifera* to *cecropia* on one axis and between *adami* and *intermissa* on the other (Fig. 13.1).

   Of the 13 Mediterranean races discriminated, four belong to the oriental group (no. 1 *anatoliaca*, no. 2 *adami*, no. 3 *cypria*, no. 4 *syriaca*) and one to the subsaharan group (no. 8 *lamarckii*). The remaining eight races can be classified according to geographic aspects in two groups:

a) Western Mediterranean
   1. North Africa
      no. 16. *A. m. sahariensis*
      no. 17. *A. m. intermissa*
   2. West and North Europe
      no. 18. *A. m. iberica*
      no. 19. *A. m. mellifera*

b) Central and Northeastern Mediterranean (South and Southeast Europe)
   no. 20. *A. m. sicula*
   no. 21. *A. m. ligustica*
   no. 22. *A. m. cecropia*
   no. 23. *A. m. macedonica*
   no. 24. *A. m. carnica*

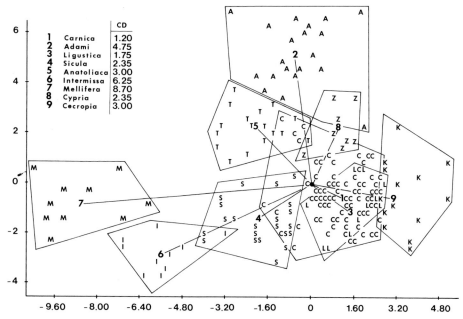

Fig. 13.1 Discriminant analysis of nine Mediterranean subspecies. N = 121. *CD* canonical distances from centroid of group

The legend in the figure:

		CD
1	Carnica	1.20
2	Adami	4.75
3	Ligustica	1.75
4	Sicula	2.35
5	Anatoliaca	3.00
6	Intermissa	6.25
7	Mellifera	8.70
8	Cypria	2.35
9	Cecropia	3.00

13.2 Statistical Characteristics of Mediterranean Races, Group (a)

Variables extracted by stepwise discriminant analysis (N = 102)

No of char.	Character	No of char.	Character
1	Length of hair (t5)	24	Wing ven. angle E9
12	Wax plate st3, long.	25	Wing ven. angle G18
16	Sternite 6, transv.	27	Wing ven. angle I16
17	Fore wing, length	30	Wing ven. angle N23
19	Cubital vein a	33	Color t3
		35	Color scutellum

Characteristics of Races

Race	Canonical variable	
	Factor 1	Factor 2
12 *adansonii*	−8.10	−0.29
16 *sahariensis*	−3.42	−2.23
17 *intermissa*	4.74	1.33
18 *iberica*	6.55	−2.57
19 *mellifera*	5.21	−4.04
20 *sicula*	4.17	2.89

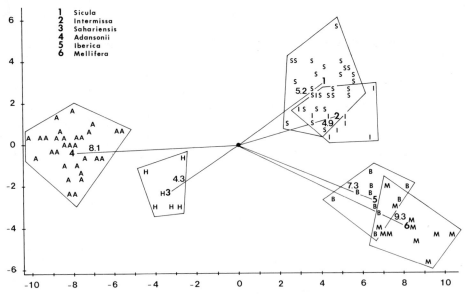

Fig. 13.2 Discriminant analysis of the West Mediterranean *mellifera* subspecies together with *sahariensis* and *adansonii*. N = 102

The clusters are plotted in Fig. 13.2: *sicula* overlaps in this analysis with *intermissa*, *iberica* with *mellifera*. If the centroid of the *intermissa* cluster is taken as center and not the 0-point of the coordinates, three phenetic trends can be observed: (Fig. 10.6) (a) *iberica- mellifera*, (b) *sahariensis- adansonii* (and other subsaharan races), (c) *sicula- carnica- cecropia* (the latter two shown in Fig. 13.1). Close links exist across the Mediterranean; morphometrically, *sicula* and *iberica* could as well be grouped with N African races as with South European and West Mediterranean races.

13.3 North Africa

13.3.1 General

Africa north of the great deserts, extending from the Red Sea to the Atlantic, is known for its climatic contrasts: extreme heat during the day and winter temperatures well below the freezing point in the interior at night (Walter 1958); low rainfall, varying from year to year. All organisms of this region have to dispose of special adaptations to survive these hard conditions, and *Apis mellifera* shows its extreme potential there. If bees of the north surprise by their capacity to survive more than 6 months of cold weather without flight this is surpassed in some respect by long periods of drought when the food supply may be limited to several weeks out of a whole year.

Somewhat mitigated are the conditions on the Mediterranean coast with higher humidity and manifold nectar sources. Egypt takes a position of its own as far as vegetation and honeybees are concerned. With even less precipitation than the other inhabited parts of the region (200 mm in the delta to less than 25 mm south of Cairo, Grünewald et al. 1983), and with similar temperatures, the Nile valley is irrigated, allowing for plentiful year-round agricultural vegetation. The local bee *A. m. lamarckii* differs basically from the predominant race of the region, *A. m. intermissa*, and it is included in tropical subsaharan races for reasons of morphometric classification. It takes a clear intermediate position between the oriental group of races (*syriaca*), the tropical African and North African races (*sahariensis*; Fig. 12.5). Since the Egyptian bee conforms to east Mediterranean races in behavior, it is a matter of convenience to discuss *A. m. lamarckii* in connection with those of North or tropical Africa.

North African races function also elsewhere as intermediates: via *sicula* to the races north of the Mediterranean (*macedonica, carnica*), via *iberica* to the West European *A. m. mellifera* (Figs. 13.1, 13.2) and finally, via *adansonii* and *scutellata* to the subsaharan races. *Sahariensis* takes a central place in the complex system of phenetic similarities. It is distinctly different in many characters from its immediate neighbor, *intermissa*, and not just a "color polymorph" as was concluded from a too scanty selection of characters (DuPraw 1964). Now restricted to a narrow strip in S Morocco and W Algeria, its multiple relation to races far south and east (even as far as *syriaca*) suggest a formerly much wider area of distribution.

The likely significant role of *A. m. sahariensis* in the evolution of Mediterranean honeybee races can be discussed only by considering the unsteady climatic history of this region with repeated periods of aridity and relative humidity since the Miocene. Most important for the distribution of the present fauna was the "Holocene pluvial" (period of the middle terrace, 15,000–8000 years B.P. with rivers, grassland and Mediterranean or temperate zone trees (*Alnus, Tilia, Quercus, Olea, Myrtus, Erica*) in the central Sahara. The zone of desert was reduced to narrow strips 8000 years B.P. (Fig. 13.3). In the neolithic period one of the first human cultures flourished there, including cattle raising, and producing marvellous rock paintings. *A. m. sahariensis* is probably a relic of this period with a formerly much wider distribution, as are a few cypress trees of an endemic species, fishes and crocodiles in "gueltas" with "fossil" water, and even a human population of the central Sahara (Gabriel 1978; Sonntag et al. 1976; Williams and Faure 1980; Schwarzbach 1974).

The relation between these races can be further elucidated by the distribution of allozyme alleles. In N Africa the S-allele ("slow") of Mdh is predominant by 90% while the rest is taken by allele M. The frequency of S decreases in N Africa from E to W and even more so in Spain from S to N (Fig. 4.8). In France virtually only allele M is found (Cornuet 1982 b). In Sicily the S-allele occurs together with a low frequency of the allele F (Badino et al. 1985). It would be of special interest to study enzyme polymorphism in other Mediterranean races.

In Africa bees increase in size from S to N (Chap. 4) and they also change proportions by becoming broader through the abdomen. The index st6 decreases from 84.5 in *adansonii* and 82.6 in *sahariensis* to 81.2 in *intermissa*, 81.0 in *iberica* and 78.5 in *mellifera*. All subsaharan races (including *lamarckii*) have indices of slenderness higher than 84.0, showing clear discrimination from N African races.

Instructive in regard to factors governing evolution in honeybees is a compari-

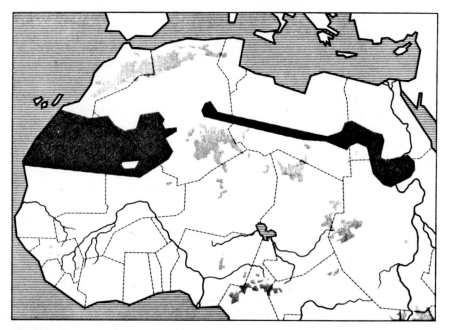

Fig. 13.3 Reconstruction of the reduced desert areas in the Sahara (more than 50 mm rain) 6500–5000 B.C. (Gabriel 1978)

son *sahariensis-yemenitica.* Both races live in similar environmental conditions of high aridity and temperatures with extreme fluctuations, short periods of activity, and comparable geographic latitude (*sahariensis* 30–32°, *yemenitica* in Saudi Arabia with its most extreme type, 25° lat.). Both races, however, show substantial differences in size, proportions, hair length, color, and wing venation. Although the phenotype of honeybees in general shows correlations to factors of environment (e.g., geographic latitude and altitude), it is clearly also the result of their individual history.

13.3.2 Description and Natural History of Local Races

13.3.2.1 Apis mellifera sahariensis Baldensperger (1922):61

The well-known apicultural pioneer Philipp I. Baldensperger made a long strenuous trip on horseback through W Algeria to S Morocco in 1921 to try to verify a rumor about a special gentle, yellow bee in the oases of this region. He returned to France with a nucleus of yellow bees which he named *A.m.sahariensis.* He, as well as others, reported the following locations for this bee: along the Djebel Amour and Ain Sefra in Algeria, and various oases from Figuig to Ouarzazate in S Morocco (Baldensperger 1922, 1924, 1932; Haccour 1960; Br.Adam 1983). In the oases of the eastern Sahara *A.m.intermissa* is found and not *sahariensis* (Br.Adam 1983). This author collected samples of bees from several oases in Tafilalet (Morocco) during two journeys in 1962 and 1976 which provided the basis for the present morphometric analysis.

All authors were impressed by the small number of colonies encountered (frequently only abandoned hives) and the harsh climatic conditions with temperatures ranging from $-8°$ to $50° C$. "We found isolated miniature pockets of a yellow bee, wedged between the Atlas and the Sahara" (Br. Adam 1983). Feral colonies are found outside the oases, in the rock niches of the wadis (e. g,. near Zagora, J. Louveaux pers. commun.; Baldensperger 1922).

A. m. sahariensis is smaller than *intermissa* in all dimensions, with a more slender abdomen and higher CI, but it is larger and broader than *adansonii* (Table 13.1). The light bands of the first abdominal tergites are not quite as wide as in Italians or in *adansonii*, and rather light tan than yellow. The drones (n = 6 samples) are small, even smaller than those of *scutellata*, with a tan-yellow band on T_2 and large spots on T_3. In behavior *sahariensis* is very different from *intermissa*: moderate tendency to swarm, restricted number of queen cells and immediate elimination of virgins during the swarming process. Little use of propolis, weak defense reaction but nervous running on the comb when disturbed ("docile"). "There is no similarity between the two races" ($=$ *intermissa* and *sahariensis*; Br. Adam 1983).

Baldensperger (1932) recommended the Saharan bee for apiculture and Haccour raised many queens and sold them to European beekeepers. But this bee is hardly able to survive the winter there, although when hybridized with the Buckfast bee it shows an astonishing heterosis effect (wintering ability, quantity of brood, and honey yield (Br. Adam 1983)

13.3.2.2 *Apis mellifera intermissa* Buttel-Reepen (1906): 187

The "Tellian" bee of the Maghreb (the coastal, mountainous region south of the Mediterranean) provides a distinct interruption of the general pattern of geographic variation in honeybees following Bergmann's rule (decrease in size from N to S). If compared with its immediate northern neighbors, *ligustica* and *sicula*, *A. m. intermissa* is larger in almost all body parts (Tables 13.1, 14.1). Also the special rule for honeybees that southern races are less pigmented does not apply, at least as far as *ligustica* is concerned. However, a gradual increase in size is observed in the West (Fig. 4.12), from N Africa to Iberia with a bee belonging to the same chain of races, thus confirming the restriction of Bergmann's rule, i. e., that it applies only to variation within the same species, or as in this case, the same "Formenkreis" (Rensch 1929).

Intermissa is one of the "big" *mellifera* races with a large population (about 1 mio colonies) located between the Atlas and the Mediterranean or Atlantic coast, extending across 2500 km. Buttel-Reepen (1906) made a correct guess with the name *"intermissa"*, indicating the intermediate position between tropical African and European races. As can be shown by multivariate analysis (Fig. 13.1), *intermissa* phenetically takes the position of a link to races in the north (*sicula*, *iberica*) as well as in the south (*sahariensis* which for its part is a link to races in the south and east (Fig. 10.6). The closest morphometric relationship exists with *sicula*, with overlapping clusters in several of the analyses (Fig. 13.2).

Table 13.1 Characteristics of honeybee races from North Africa and West and North Europe (Means and standard deviation of n samples; measurements in mm)

Character	Sahariensis n = 6	Intermissa n = 20	Iberica n = 17	Mellifera n = 10
Terg 3 + 4	4.100	4.433	4.562	4.640
sd	0.156	0.159	0.161	0.121
Proboscis	6.242	6.381	6.443	6.052
sd	0.230	0.212	0.221	0.147
Fore wing length	8.948	9.185	9.253	9.334
sd	0.165	0.241	0.173	0.111
Hind leg	7.696	8.023	8.288	8.099
sd	0.179	0.201	0.186	0.169
Distance wax plates	0.232	0.275	0.261	0.225
sd	0.055	0.065	0.051	0.047
Stern 6 Ind	82.84	81.16	81.03	78.61
sd	2.42	3.41	3.48	4.66
Cub Ind	2.62	2.33	1.84	1.84
sd	0.41	0.36	0.27	0.28
Hair length	0.230	0.271	0.257	0.436
sd	0.026	0.054	0.046	0.055
Color terg 3	6.78	3.27	2.53	3.45
sd	2.69	2.17	1.56	1.02

Some morphometric characteristics of intermissa are presented in Table 13.1. The most conspicuous traits are the shiny black pigmentation of the exoskeleton (including scutellum), almost completely exposed because of the scarce hair cover which may completely disappear in older bees, and the smaller body size compared to the dark European races (*mellifera* and *carnica*). Surveying the whole set of biometric data, it can be observed that this bee is intermediate in every respect without substantial deviations in single measurements or indices. Drones (n = 6) are as dark as in *A. m. mellifera*, but smaller (Fig. 10.8).

The behavior of *A. m. intermissa* has become known in the apicultural world since the early days of modern beekeeping through the export of queens and intensive and productive beekeeping in north African countries (Buttel-Reepen 1906; Abushady 1919; Baldensperger 1924; Barbier 1969; Hicheri and Bouderbala 1969; Ruttner 1980; Br. Adam 1983). Brood cycle and reproductive behavior are shaped by the exceptional seasonal and periodical contrasts of climate. The brood cycle corresponds exactly to the floral cycle: abundant spring nectar flow, summer heat and drought (May-September) with a complete brood stop and a relatively humid autumn and winter with a modest nectar supply (eucalyptus, rosemary and other wild plants) stimulating a second peak of brood activity. This constitutes the typical Mediterranean bimodal type of brood rhythm in its most extreme form (Fig. 10.4). Races of the temperate zone, even if of highly reputed origin, are not able to adapt themselves, and dwindle within a short time. Importation of foreign stock has not been successful in Libya, Tunisia, and Algeria (pers. information from different sources in these countries). Tellian colonies brought to England by Br. Adam maintained their enormous brood-rearing activity even late in autumn when all other colonies were broodless.

Usually more than 100 queen cells are constructed during the swarming season (Fig. 13.4) and many virgin queens stay alive until one queen is mated as is observed in other Mediterranean races (Chap. 4). Afterswarms are frequent. One colony in the oasis of Laghouat (Algeria) produced seven swarms during one season, until only a few hundred bees were left in the mother colony. Nevertheless, the colony survived and developed to full strength in the next season showing the enormous vitality of this bee (Br. Adam 1983). This extremely high reproduction rate in favorable years copes with heavy losses (up to 80% of the population) caused by periods of drought sometimes lasting for several years. After dequeening, laying workers appear within a few days, the shortest interval found in seven races tested (Ruttner and Hesse 1979). The claim of the commercial beekeeper Hewitt, made almost 100 years ago, that worker bees produce female offspring as in the Cape bee, which is quoted in textbooks again and again, has never been confirmed (Baldensperger 1924; Br. Adam 1983; own inquiries).

The characteristics least appreciated by beekeepers are strong defense behavior, extreme nervousness ("when a hive is opened the bees boil over and mill around in a most alarming manner"; Br. Adam 1983) and abundant use of propolis. Brood diseases seem to be more widespread than in other races. There exist, however, remarkable differences in these behavioral traits within the population. Otherwise *A. m. intermissa* appears to be rather uniform in morphology and behavior throughout its territory, although it seems possible to discriminate subpopula-

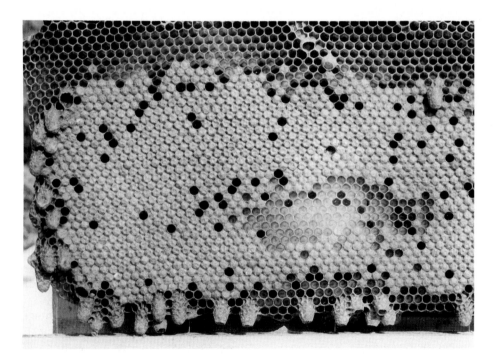

Fig. 13.4 Brood comb of an *intermissa* colony ready to swarm. Twenty five queen cells are visible on one side of the comb

tions (Tunisia, Algeria, Morocco) by multivariate analysis. Our data are too limited to reveal more than an indication.

There is one exception: the largest bee found so far, detected in a very restricted area of the Rif mountains in the middle of the *intermissa* territory. It was described because of its astonishing body dimensions as *Apis m. major* (Ruttner 1975). The values by far exceed those of the region: length of fore wing 9.52 mm, hind leg 8.42 mm, tongue 7.00–7.12 mm. The drones do not match the extreme size of the workers. Although they are distinctly larger than *intermissa*, they are smaller than those of *mellifera* (wing length 12.08 vs. 12.24 mm). Br. Adam visited the location (between Tetuan and El Hoceima) in 1976 and took some queens to England. He found no difference in behavior from *intermissa* (Br. Adam 1983). It would be of interest to investigate the distribution and possible origin of this giant variety and to consider further the most appropriate taxonomic status.

13.4 West and North Europa

This is a region of a rather uniform type of bees over a large area with little radiation, quite in contrast to the variegated Central and East Mediterranean. This difference may be explained by the relatively late colonization of the territory between the Pyrenees and the Ural after the last glaciation. Historically, both races described in this section originate from the Meditarranean coast.

This uniform type constitutes the westernmost branch of *A. mellifera*, extending from N Africa through western Europe to Scandinavia and north of the Alps as far east as the Ural mountains. It reaches to the northern limits of *Apis*, with the largest bees of *A. mellifera*. In S-N direction an overall increase in size and in width of the abdomen is observed. The graded variation is interrupted by barriers, the Mediterranean and the Pyrenees, separating *iberica* from *intermissa* and *mellifera* (stepwise cline; Huxley 1939). A number of obvious traits are common to this type, e.g., narrow tomenta, nervous and irritable behavior and abundant use of propolis. The more important ability to survive long humid and cold winters increases gradually from subtropical Africa to cold-temperate N-Europe

13.4.1 Apis mellifera iberica Goetze (1964):25

The honeybee of the Iberian Peninsula is something of a stepdaughter in the large family of honeybees of Mediterranean origin in spite of the size of its population and the economic importance of apiculture in the region. Named by Goetze in 1964, it was never properly described. There is only one detailed report on this bee by Br. Adam, who traveled through the whole peninsula in September/October 1959. Therefore, only limited knowledge about the behavior but a considerable collection of samples is available.

In this link between N African and N European honeybees several important changes occurred compared to the Tellian bee (Table 13.1): increase in overall size, change of proportions (fore wings becoming narrower, abdomen broader) and decrease in CI, with unchanged hair length. In behavior the Iberian bee could almost be taken to be a variety of *intermissa*: quick defense reaction, nervousnees on the comb, propensity to swarm, and ample use of propolis. However, there are evidently also temperate-zone features: nothing is said about polygyny during the swarming period; the cold and long winters in the interior are survived without difficulty. The definite proof of its "temperate" character became evident when transferred to the tropics of the New World: while *mellifera* and *ligustica* colonized temperate North America with feral populations immediately after importation, *iberica* remained in need of shelter provided by men for centuries in tropical South America. No feral colonies of this race were found (Taylor 1985). As soon as a tropical bee was imported from Africa, a large feral population rapidly developed in the forests there.

On the Iberian Peninsula more than 1.6 mio. bee colonies were already counted in 1959, more than 6 hives per km². In the interior there are great climatic contrasts between the seasons and the years which require highest vitality and resistance. "The bee must either adjust itself to the environment or perish" (Br. Adam 1983). Our data provide a general survey of the Iberian bees. Recently Santiago et al. (1986) showed that in regions with different climate in NW Spain two different types of *iberica* are found. The two ecotypes are separated by the Cantabrian cordillera.

13.4.2 *Apis mellifera mellifera* Linnaeus (1758):576

The "common European black bee" was *the* honeybee until a few decades ago, the holotype of Linnaeus' *A. mellifera* of 1758 and "*A. mellifica*" of 1765 (see p. 58). Only lately has it been fully realized that this race is nothing else than a small sector of the geographic variation of the species covering only a part of the huge total territory. The major part of knowledge of the morphology and biology of honeybees in Europe before World War II was achieved with this bee. The extensive bibliography is summarized in Berlepsch (1873); Buttel-Reepen (1906); Alpatov (1929); Butler (1954); Goetze (1964); Br. Adam (1983), and Ruttner (1987).

Two unique characteristics are significant of this race:
- length of the abdominal cover hair, up to 0.50 in the north (= double the length of bees from a warmer climate).
- large body size with a broad abdomen.

These two characters together with the narrow tomenta and dark pigmentation give this bee an unmistakable velvet-black, heavy look (Fig. 13.5). The taxonomic diagnosis is complemented by a CI lower than 1.85. Several biometric data are given in Table 13.1. The drones are completely black (in some strains including the posterior rim of the tergites) and together with *caucasica* the largest of all races investigated (n = 7 samples, originating from regions with very different climates – S France, England, Norway, Bashkiria): length of fore wing 12.24 mm, width 3.92 mm, hind leg 10.10 mm (Table 11.4). Specific is a very low CI $\bar{x} = 1.425$) which can be below 1.0 in some samples. In the phenogram of a factor analysis the *mellifera* samples (workers as well as drones) take an extreme, compact position in the M-branch of the "Y" (Figs. 10.5, 10.6, 10.7).

The original area of *mellifera* covers all of France, the British Islands up to Scotland and Ireland, Central Europe north of the Alps and the plains of N

Fig. 13.5 *A.m.:nellifera* from the North Alps (Tyrol, Austria)

Poland, and the USSR east to the Ural mountains. The island of Corsica has a black bee population greatly overlapping the *mellifera* population in a factor analysis (Battesti 1980). In Scandinavia, honeybees originally were restricted to S Sweden, to the northern limits of trees sensitive to climate like lime and hazel *(Tilia, Corylus)*. No bees were kept in Norway before the 19th century (Lunder, pers. commun.). At present, in Norway bees are kept for economic purposes up to 65° latitude (250 km N of Drontheim). This shows that not the length of winter, but the lack of suitable nesting sites is the limiting factor of distribution (there is a rich supply of nectar from heather in this region).

In Russia, the northern boundary is marked by a line where rivers are frozen for less than 6 months (Fig.10.3; Alpatov 1976). That is 60° lat. at the Baltic Sea (Leningrad) and 57° near the Ural. It is this large forest region between the Oder river and the Ural which provided most of the honey and wax for the European market in medieval times. Bashkiria in the southern Ural is famous for its lime forests and the considerable populations of feral bee colonies and bears – in spite of winter temperatures as low as $-45°$ C. Farther south where the forest changes to the Ukrainean steppe, a gradual transition occurs to the Ukrainean Bee (p.51).

Synonyms. A large and very dark type in the heather region of N Germany and the Netherlands was named *A.m.lehzeni* (Buttel-Reepen 1906). A high swarming propensity (regular afterswarms which may swarm again in the same season, drone production by young queens) was recorded as peculiarity of this variety, correlated with a late honey flow from ling (Berlepsch 1872). *A.m.mellifera* from Middle Russia is sometimes quoted as *A.m. silvarum* (forest bee; Alpatov 1935b).

In behavior, *mellifera* has many traits in common with the west Mediterranean group *intermissa- iberica* (p.231; Br.Adam 1983): nervousness on the comb, easily released defense behavior, ample use of propolis, high rate of drifting. Typically, a flat seasonal brood rhythm is observed: Slow increase of brood quantity in spring, late flat peak, slow decline in autumn (Fig.10.4). This is an adaptation to the heather nectar flow late in summer, found all along the Atlantic coast from Portugal to Norway ("Atlantic" type of brood rhythm in contrast to the "continental" type of *A.m.carnica*).

Ecotypes. Differences in the seasonal brood rhythm were studied in various regions of France. At the south coast and the island of Corsica the Mediterranean climate is prevailing. During summer the brood production is reduced or completely abolished, while in autumn a second peak is observed. In the heather region of the SW (Les Landes) the main brood activity is shifted to summer and early autumn, while in the region of Paris, with no nectar flow in summer at all, a clear spring-type brood activity with one early peak is found. The exchange of colonies between Paris and Landes showed that the original pattern of brood activity is maintained in the new environment, resulting in disparity with the floral cycle and, consequently, poor performance (Louveaux et al. 1966; Louveaux 1969). These ecotypes can also be discriminated morphometrically – significantly just by the newly acquired characters of the race, length of hair, and distance A of the cubital vein (Cornuet et al. 1978). This demonstrates the high plasticity of a race which was able to colonize a huge area only relatively recently and to adapt to climates as different as the humid, moderate Atlantic coast to the high valleys of the Alps and the extremely continental Ural region with seasonal extremes of temperature. However, this race evidently did not succeed in traversing these mountains, since *mellifera* bees were brought to Western Siberia by colonists only in the beginning of the 19th century, while East Siberia (Ussuria) was colonized with Ukrainian bees by the end of the same century (Alpatov 1948, 1976).

Probably still other ecotypes could be detected, but by now *mellifera* is heavily hybridized or has even completely vanished in many countries, especially in Central Europe. The vitality of *mellifera* in cold temperate forest regions is shown by the speed in colonizing N America with feral colonies. In Tasmania (43° S) thousands of colonies of the imported black English bee live in trees and cliffs and they are found to be more "typical" than the present descendants of their common ancestors in Britain (Ruttner 1978).

CHAPTER 14

Honeybees of the Central Mediterranean and Southeastern Europe

14.1 General

The "honeybee province" described in this subchapter comprises a group of closely related geographic races and subraces. Some of them were well established during the last century, while others had a disputable taxonomic status. Overlapping of phenetic types occurs partly as a result of the relatively small extension of the area, a relatively recent divergence, and lack of geographic isolation. The geological age of the populations is certainly not without significance. Only in the southern part it may reach back as far as the Pleistocene. Only in the north, however, honeybees do not exist longer than the Atlanticum (8000–10,000 years). The bee population of the Ukrainian steppe is of recent origin (about 500 years; Skorikov 1929b).

The region comprises the Apennine and Balkan Peninsula with the Alps as northern, and Sicily and the West Aegean islands as the southern boundaries (Fig. 14.1). The east European plains in Poland and the USSR constitute an area of gradual change from the Balkan bees to the northern *A. m. mellifera*. The "Ukrainian bee" as mentioned by Russian authors (see Alpatov 1929) morphometrically clearly belongs to a Mediterranean race, *A. m. macedonica*. The climate of this whole region, which extends between 36–51° N lat., ranges between cold temperate (mainly continental in the N and E) and warm temperate (mediterranean) on the coast.

The bees of the central Mediterranean and its "back yard" are well separated from the oriental group, as they are from the west European races. *A. m. sicula* constitutes a link with the Tellian bee, *intermissa*, in N Africa (overlapping to both sides). The taxonomic units within the group have much in common: medium to tall size, relatively long proboscis, short hair, relatively high cubital index, and predominantly dark color (except for *A. m. ligustica*, which was called the "blond cousin" of *carnica*).

All these factors contributed to an unsatisfactory situation in which a reasonable taxonomic classification has not been achieved. The principal question was: how many taxonomic units are found in this region comparable to other races in diversity and uniformity?

Various and repeated multivariate analyses were carried out with a total of 278 samples originating from the following countries: CSSR – 6 samples, Austria – 22, Hungary – 13, Yugoslavia – 41, Romania – 12, USSR – 6, Bulgaria – 16, Greece – 49, Island of Crete – 24, Italy – 60, Anatolia – 25. The samples were classified as preliminary taxonomic units by PCA and then introduced into a stepwise discri-

Fig. 14.1 Distribution of honeybee races in S and SE Europe

minant analysis. The result is shown in Fig. 14.2. From the eight preliminary groups, Bulgaria and Northern Greece are almost completely overlapping and were united in one taxon, which includes also the samples from the Romanian Black Sea coast and from the Ukraine. The two Asiatic clusters used as reference (West and Central Anatolia) are fairly distant from the Balkan group. *A. m. ligustica* from continental Italy takes the same distance. Four relatively compact clusters are very close together: *sicula*, *macedonica* (Bulgaria and N Greece), *cecropia* (S Greece) and *carnica* (Austria, CSSR, Hungary, Yugoslavia, Romania).

Morphometric data are given in Table 14.1 and details will be discussed in the paragraphs on the respective races. A few odd observations, however, have to be mentioned here.

The morphometic distance between *ligustica* and *sicula* is about the same as *ligustica* and the Balkan races, in spite of the scanty georgaphic isolation of both races across the strait of Messina, while the clusters *sicula-cecropia* (with great

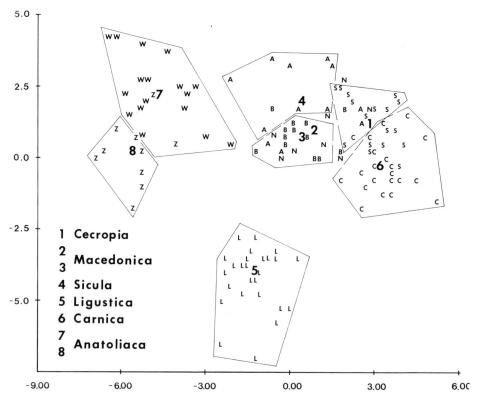

Fig. 14.2 DA of samples from Italy and the countries of SE Europe *2* Northern Greece, *3* Bulgaria, *7* West Anatolia, *8* Central Anatolia

geographic distance) are overlapping. The distinctness of *sicula-ligustica* is confirmed by the distribution of alleles in the polymorphic enzyme loci MDH 1 and Est (Fig. 4.8). In Sicily the alleles MDH 1-F and Est-S are frequent or fixed, while they are rare or missing on the Italian Peninsula (Bsadino et al. 1985).

In this whole group of races no relation seems to exist between geographic and morphometric distance. As far as the morphometric distance is concerned, Sicily is closer to Bulgaria than to Italy. This situation makes it difficult to speculate regarding the routes of migration.

As a result of numerous multivariate analyses, the following five honeybee races of the Central Mediterranean can be discriminated:

20. *A. m. sicula* Montagano
21. *A. m. ligustica* Spinola
22. *A. m. cecropia* Kiesewetter
23. *A. m. macedonica* Ruttner
24. *A. m. carnica* Pollmann

The Balkan and Apennine group of races should not be considered in isolation, but together with its neighbors. Including the race of the opposite Mediterranean coast, *A. m. intermissa,* from North Africa, into the morphometric analysis, it becomes immediately evident that a close similarity exists with *A. m. sicula* and, via this race, with *A. m. cecropia*. It has to be borne in mind that this part of the Mediterranean Sea is very shallow and fell dry to a large extent during the glacial depression of the sea level, dividing the Mediterranean into a western and eastern basin. Only narrow straits of water were left dividing Africa from Europe and Apulia from southern Greece. The geomorphological situation during the late Pleistocene explains the existence of a *Central Mediterranean gene center* extending from North Africa to South Greece across 1800 km between the 36th and 39th degree of latitude. All honeybee races found in Europe can be traced back to this center. Morphometrically and phylogenetically *A. m. intermissa* belongs to this Central Mediterranean group, but for reasons of tradition it has been dealt with together with North African or West Mediterranean races.

14.1.1 Statistical Characteristics

The races of this group can be discriminated by using the following 13 characteristics for a discriminant analysis:

No of char.	Character	Coordinates Race	Factor 1	Factor 2
5	Femur	*carnica*	3.15	−0.29
6	Tibia	*macedonica*	0.69	0.96
7	Basitars. long.	*cecropia*	2.89	1.15
16	Stern. 6, trans.	*sicula*	0.39	1.92
17	Fore wing, length	*ligustica*	−1.15	−4.08
19	Cub. vein b	(W Anatolia)	−4.67	2.35
21	Ven. angle A4	(Central Anatolia)	−5.53	0.18
24	Ven. angle E9			
25	Ven. angle G18			
26	Ven. angle J10			
29	Ven. angle L13			
31	Ven. angle O26			
33	Color T3			

14.2 Characteristics and Natural History of Local Races

14.2.1 Apis mellifera sicula Montagano (1911):26

The first author describing the Sicilian honeybee is universally assumed to be B. Grassi (1881). However, reading the original publication "Saggio di una monografia delle api d'Italia", it is evident that Grassi speaks generally of "l'ape sicula", and not of a taxonomic unit. He discriminates several types; the dark variety, which is compared with the German black bee, is believed possibly to be a hybrid between the native yellow and the cecropian bee. The first clear description of a special race with the scientific name "*Apis sicula*" dates from Montagano (1911).

Sicily is separated by a narrow strait from the continent, providing only an incomplete isolation for bees. Additional *ligustica* queens are permanently brought to Sicily in great number (Longo 1984). Nevertheless, the 30 samples examined represent a relatively homogenous cluster (Fig. 14.2); this may have different reasons: (1) The Sicilian bee is a population with its own history and different from Italians through specific adaptations to a subtropical climate with a hot and dry summer season and special enemies. (2) A large population of more than 100,000 colonies 90% of which were kept in some provinces (Syracuse, Ragusa) up to 1980 in traditional "ferula" hives (Longo 1980, 1982, 1984). (3) Samples were collected in Sicily on a large before the start of modern apiculture scale, by Br. Adam in 1952 and by M. Alber in 1971/72. The latter may have been inclined to select samples for "purity", that means he avoided yellow colonies. Thanks to these two apiculturists, bees from all over Sicily, including the hard-to-reach interior, were obtained for analysis, and it can be stated that at the time of collection unhybridized colonies of *sicula* were found in all parts of the island.

At first sight, *A. m. sicula* always was and still is recognized by its dark color. The mean value of color classes of tergite 3 in our samples varied from 1.4 to 4.3, that is from almost completely dark to large yellow spots (but no bands; Figs. 6.10, 6.11). The hair of the thorax of workers and drones is not grey or brown as in other dark races, but yellowish.

Of course, morphometrically *A. m. sicula* has to be compared with the neighboring *A. m. ligustica*. Both races are almost identical in body size and length of hind legs, but *sicula* has smaller (shorter and narrower) fore wings, a shorter proboscis, somewhat longer hair, and lower cubital index. The vicinity of North Africa is indicated by a broader abdomen (index S6 82.5 vs. 83.5 in *ligustica*, and 81.2 in *intermissa*).

Sicula drones are completely dark in their exoskeleton. They have distinctly shorter wings and somewhat shorter hind legs than *ligustica*.

The taxonomic individuality of *A. m. sicula* is confirmed by allele frequencies of polymorphic enzyme loci. In the MDH-1 locus the allele F ranges from 75 to 100% in Sicily, while the S allele is rare; in northern and central Italy the frequencies are the reverse. In the Est locus, the allele S prevails in Sicily, which is not found on the Italian peninsula (Badino et al. 1984; Fig. 4.8).

A remarkable geographic variability *within* the Sicilian population was noted already by early investigators (Grassi 1871; Alber in Longo 1984). Multivariate analysis reveals four geographic groups, although with partial overlapping: NE (Messina) - NW (Palermo) - central region (Nicosia) - SE (Syracuse - Ragusa). Size is diminishing in the order listed. This corresponds to the discrimination by Alber (Longo 1984) of larger bees in the NW and smaller ones in the SE; but his idea that this could be the consequence of different importations (from Greece to the SE and from N Africa and Spain to the NW) is certainly quite hypothetical.

Only a few published reports are available on the biology of *A. m. sicula* although this is a field of special interest. Therefore, a number of questions were put to professional and amateur beekeepers with the kind assistance of M. Lodesani (Bologna) and S. Longo (Catania). The information received from 18 beekeepers (owning from 20 to 3000 colonies, a total of 16,000 colonies), supplemented by the profound knowledge of Dr. Longo, resulted in an almost unanimous description of the biology of *A. m. sicula*: rather gentle and quiet behavior when manipulated, low tendency to follow the apiculturist after being disturbed. High (white) capping of honey. Ample use of propolis late in summer and autumn. Brood rearing and maintenance of drones almost the year round except for a very short period in November/December. Very strongly inclined to swarm, especially in the traditional horizontal "ferula" hives (volume 32 l). Division of traditional colonies is practised in the same way as in ancient Egypt (Longo 1980). The number of queen cells during the swarming process is extraordinarily high: Genduso and Alber (1973) found an average of six colonies of 282 cells (from 64 to 430 cells) and the interviewed beekeepers reported "100 or more queen cells".

The most surprising, but generally confirmed, observation is that no swarm departs until a number of virgins have hatched. Mother and daughters live peacefully together in the colony before the swarm leaves. This behavior resembles that of the Syrian and the Egyptian bee. In this respect *A. m. sicula* is completely isolated among European races.

Another trait of some tropical bees (found also in *A. cerana*, p. 136) is the defense strategy used against the dangerous predator *Vespa orientalis*. While *ligustica* workers are killed in great numbers when actively defending their hive by attacking the hornets in flight, *sicula* bees stop flying completely during the hot hours of the day (when hornets are most active) and reduce the entrance by using propolis.

One of the traditional regions for beekeeping is the SE corner of Sicily, the Monti Iblei between Syracuse and Ragusa. Within this region 70,000 colonies are regularly transported according to a traditional scheme: early spring flow at the coast, thyme bloom in summer in the highland and bloom of the carob tree in autumn. The harvest is relatively modest compared to this activity – 2–4 kg per traditional colony and 25–30 kg per Langstroth colony (Longo 1982).

14.2.2 Apis mellifera ligustica Spinola (1806):35

The Italian honeybee is not only the most generally distributed race in the world but also the reason for two important discoveries:

1. After Johannes Dzierzon had imported Italian colonies from the region of Venice to Germany in 1852, he observed that the locally mated daughters of these queens produced yellow hybrid workers and pure black drones. His ingenious conclusion was that drones originate from unfertilized eggs – one of the first scientific proofs of parthenogenesis in biology (Siebold 1856)
2. After their importation to the New World, the beekeepers soon preferred Italian bees to the hitherto common black NW European bees because of superior gentleness and fertility; but they soon became aware of the side effects, namely vicious and nervous hybrids. To secure pure matings of Italians, early attempts at artificial insemination have been made since the late 19th century. This finally resulted in the widespread practice and highly efficient technique of instrumental insemination (Laidlow 1987).

The Italian bee is also of outstanding importance for apiculture all over the world. Br. Adam writes (1983): " Indeed, I believe apiculture would never have made the progress it did without the Italian bee". The reason for this is adaptability to a wide range of climatic conditions and a combination of behavioral traits which are needed in modern apiculture: docile and quiet on the comb, if disturbed; prolific, with a tendency to build big populations and stores of honey without swarming, little use of propolis.

From the point of view of biogeography and taxonomy, fewer problems have to be discussed than with almost all other races. The area of distribution, the Apennine Peninsula, is confined by high mountains and the sea. No defined local populations of this race were so far detected.

As to exterior traits, *A. m. ligustica* was given a peculiar position among European races of honeybees because it is the only one with yellow pigmentation. However, since classification was not primarily done for color, it became evident that only small differences exist between the Italian and the bees of the Balkan: *ligustica* workers have a smaller and broader abdomen, but identical wing and leg length when compared with *carnica* from neighboring Yugoslavia. Proboscis is slightly longer, hair shorter and several venation angles are slightly different (Table 14.1)

Thus the most important, but not the sole, difference is the color of abdominal tergites and scutellum. Since "yellow" and "dark" races of the Mediterranean show no overlapping in their variation of color, while other morphometric characters do, using color together with only a few other measurements results in splitting all races in only two groups, "yellow" and "dark" (p. 40, Cornuet et al. 1975). The mean value of color of *ligustica* bees (35 samples) for tergites 2, 3, and 4 was determined to be 7.6, 7.1, and 3.8 (that is, two yellow bands and large yellow spots at tergite 4). In drones the pigmentation is similar at the same tergites. A yellow tinge of hair in workers and (still more distinctly) in drones is worth mentioning. Since the yellow pigmentation is controlled by at least seven dominant polygenes (Roberts and Mackensen 1951), any hybridization with *ligustica* is easily recognizable, although the extension of yellow areas is reduced in F_1. On the other hand, not all yellow bees are identical with *A. m. ligustica*. This is an important consideration in countries where *ligustica* was imported and hybridization occurred.

With *A. m. meda* from Iran a difficulty in the other direction arises since the very similar *ligustica* is compared with a race of the same pigmentation: if both races are analyzed jointly, together with a group of others, the *ligustica* cluster is almost invariably atomized. Yet, the two races can be canonically separated by discriminant analysis with only five characters selected: hair length, tergite 3 longitudinal, width of fore wing, the two wing venation angles G18 and I10.

In the northwest of Italy (Province Imperia, Valle d'Aosta) *A. m. mellifera* or hybrids with this race are found (Leporati et al. 1984; Marletto et al. 1984a, b). This zone of hybridization was confirmed also by a gradual shift of allele frequencies in the enzyme malate dehydrogenase (MDH-1): while *A. m. mellifera* of France is monotypic for the allele M, in Italy the two alleles S and (with lower frequency) F are found. Only in the northwest frontier zone is the allele M found in variable frequency (Badino et al. 1984; Marletto et al. 1984a). Hybridization with *carnica* is found in the NW corner of Italy (Valli et al. 1983)

In southern Italy (Calabria) the enzyme pattern changes fundamentally: The S allele of the MDH locus diminishes, the F allele (typical for Sicily) prevails and a new allele, F_1, is found (Fig. 4.8). Allele frequencies are different also in the esterase locus in Calabria (Badino et al. 1985). Unfortunately nothing is known about MDH alleles of Greece. Considering the morphometric indication of "a Central Mediterranean gene center" (p. 243), any comparison would be of the highest interest.

At any rate, the bees of the south are different from those of other parts of Italy. Valli et al. (1984) state a certain influence of *A. m. sicula* in this region. "South of this region (Catanzaro) the worst possible type of mongrel extant dominates the remainder of Calabria – a heterogenous conglomeration of the yellow Italian and the black native bee of Sicily" (Br. Adam 1983).

A local strain from the mountains along the Ligurian coast between Genoa and La Spezia, the "brown Ligurian bee", is repeatedly quoted by Br. Adam (1983) as the "original Ligurian bee". However, morphometric as well as enzyme investigation shows a high level of homogeneity throughout Italy, except for the few marginal areas described above.

Italian bees which are exported to other countries originate almost entirely from the efficient, traditional queen-rearing establishments around Bologna. Beekeepers tend to prefer the yellow strains; therefore Italians from the USA show more yellow color than original *A. m. ligustica* (Alpatov 1929). The bee of Sardinia seems to belong to *A. m. ligustica*, as far as can be concluded from the preliminary papers of Vecchi and Giavarini (1938) and Prota (1976).

Average *ligustica* drones are of the same size as *carnica* drones, but somewhat smaller than *mellifera*. A broad yellow band is found on tergite 2 (class 8.15) and a narrow one at tergite 3 (5.3). The hair on the thorax is distinctly yellow. The cubital index is relatively elevated (2.14).

Behavior

In contrast to the generally slight differences between *ligustica* and *carnica*, the dissimilarities in behavior are striking: while *carnica* workers are long-distance foragers with a very smooth frequency curve of waggle runs, *ligustica* starts these already at a distance of 20 m and the frequency of runs diminishes quickly (Fig. 4.9). Therefore, only very slight differences remain to indicate various distances farther away from the hive (Frisch 1967). Probably correlated with this "short-distance foraging" is the high tendency to robbing. The first to start robbing other colonies are usually yellow Italians with more attentive scout bees close to their hive. A certain type of communication dance occurs in *ligustica*, not observed in other races: the "sickle dance" (Baltzer 1956), a transitory pattern between round and waggle dance, indicating medium distances. The runs are not completed but turns are made before closing the circles, resembling the "directed round dance" of *A. cerana* (p. 141).

Different also is the learning ability regarding orientation marks: *carnica* orients itself (while foraging or homing) primarily by structures and the relative position of the goal (food source or hive entrance), color being used only secondarily. *Ligustica*, on the other hand, orients itself primarily by color (Lauer and Lindauer 1973). Consequently, there is a striking difference in drifting from one colony to the other. If colonies of different races are positioned close together yellow *ligustica* bees are soon found in all of them.

The brood rhythm of *ligustica* starts slowly in spring and comes to a peak which lasts long into late summer and autumn, irrespective of nectar flow. Winter brood is common with a short pause only by the end of the year. This rhythm of brood activity reflects the warm-temperate, yet moderate type of Italian climate. The mediterranean summer drought, less pronounced in the north and the mountains of Italy than in the mediterranean east and south, allows for continuous brood rearing during summer. Surprising is the wintering ability of Italians, even if they are far north of their original area. They are somewhat hampered by inopportune brood rearing in winter, necessitating high food consumption and resulting in an exhausted population in spring (Br. Adam 1983). However, selected strains with adapted management methods are overwintered with best results even in Finland. Swarming tendency is very low in this race and only a few swarm cells are constructed.

14.2.3 *Apis mellifera cecropia* Kiesenwetter (1860):315

Two colonies of honeybees were brought from the monastery Caesarea on Mount Hymettos to Dresden in 1860. The bees were described and named in the same year by the entomologist E. Kiesenwetter. The name "cecropian bee" was used by Virgil in his poem "Georgics" (Pollmann 1889). For a short while the *cecropia* was glorified by prominent apiculturists because of the famous "Hymettos honey", but soon it was found that its frequent rusty-brown tergite rings compare unfavorably with the "beautiful" bright yellow of Italians and Egyptians. *Cecropia* was supposed to be a hybrid between one of the yellow races and the dark north Europeans, and Buttel-Reepen (1906) eliminated it as taxon from his list. Maa (1953) keeps *cecropia* as "natio" of *A. m. mellifera*. The validity of the taxon was now confirmed by multivariate analysis (Figs. 10.5, 14.2).

The northern limit of *cecropia* was found along the line Ioanina – Metsovo – Kalambaka. Farther north, in Grevena – Kosani – Yiannitza, only *macedonica* occurred. No samples were analyzed from the east of the peninsula (Thessalia) from where similar bees as in the north were recorded (Br. Adam 1983; Ifantidis 1979 a, b). *Cecropia* covers all of the south of Greece including the Peloponnes.

Surprisingly, the bee population of the Ionian Islands, situated very close to the Greek west coast, appears morphometrically as a *carnica* subpopulation (somewhat similar to the strain of Dalmatia), not *cecropia*.

Cecropia is larger than *macedonica*, corresponding in overall size (abdomen, fore wing) to *carnica*, but it has longer hind legs and proboscis. This is another example opposed to the so-called rule that southern races are smaller. One of the most remarkable traits of this bee is the cubital index. The mean value of 3.13 is the highest of all *mellifera* races, close to some *cerana* samples. Single *cecropia* samples even measured as much as 3.45 (Ioanina, Epirus) and 3.60 (Dafni, Attica). Other characteristics are given in Table 14.1. No information was obtained about specific behavioral traits of the honeybee of southern Greece.

Table 14.1 Selected characteristics of S and SE European races compared to *A. m. intermissa* (means and standard devation; data on size in mm)

	Intermissa	Sicula	Ligustica	Cecropia	Macedonica	Carnica A
n	20	10	35	18	20	21
Proboscis	6.381	6.254	6.359	6.561	6.445	6.396
s.d.	0.212	0.257	0.126	0.177	0.140	0.154
Hind leg	8.023	7.951	7.969	8.126	8.008	8.102
s.d.	0.201	0.226	0.177	0.177	0.145	0.165
T3+4	4.433	4.376	4.348	4.455	4.466	4.514
s.d.	0.149	0.136	0.148	0.114	0.137	0.180
Fore wing length	9.185	8.976	9.208	9.184	9.180	9.403
s.d.	0.241	0.158	0.175	0.152	0.151	0.150
Vein b	1.920	1.903	1.866	1.602	1.829	1.882
s.d.	0.224	0.225	0.334	0.217	0.218	0.216
Cub Index	2.325	2.467	2.551	3.109	2.591	2.589
s.d.	0.358	0.420	0.410	0.566	0.412	0.418
Stern 6 Index	81.16	82.80	83.48	85.84	86.40	83.46
s.d.	3.41	3.51	3.31	2.70	3.76	2.91
Angle E9	19.81	21.29	23.49	22.60	22.05	23.12
s.d.	1.40	3.80	1.91	1.85	1.50	2.08
Angle G18	101.84	97.34	93.46	91.79	94.05	93.10
s.d.	3.95	3.80	2.74	4.05	3.16	3.33
Angle I10	49.24	50.54	52.13	53.87	54.81	52.20
s.d.	4.03	4.02	3.05	4.07	3.63	3.45
Angle L13	14.58	14.54	13.57	12.93	14.16	12.48
s.d.	1.61	1.28	1.49	1.42	1.58	2.96
Color terg 3	3.27	2.94	7.14	3.67	3.80	2.35
	2.17	1.21	1.25	1.01	1.49	0.74

14.2.4 *Apis mellifera macedonica* ssp. nova

Introducing the Macedonian bee as a separate geographic race is the logical consequence of a long process of research. Other than more exotic Mediterranean races from Palestine, Cyprus, and Egypt, the bees of Greece have long been ignored in spite of highly developed apiculture since antiquity. Br. Adam made one of his apicultural research travels to this country in 1952, collecting samples and making observations from the Peloponnes to Macedonia (Br. Adam 1952, 1983). Ifantidis published a biometrical survey in 1979 regarding worker bees and drones of the whole of Greece. Both researchers stated that these bees belong to the "family circle of the Carniolans" and that two populations can be discriminated, one in the north (Macedonia, Thracia) and one in the south. Finally, the multivariate analysis of samples of the whole Balkan Peninsula conducted by the author, resulted in the same discrimination of a northern and southern population of Greece. The northern population, however, was found to extend as far north and northeast as Yugoslavia, Bulgaria, Romania and the USSR (Fig. 14.1). Therefore, the prerogatives for a taxonomic unit are given – clear morphometric and geographic definition. *A. m. macedonica,* named after the kingdom of Alexander the Great (now divided between Yugoslavia, Greece and Bulgaria) is a true pontomediterranean element, as is *A. m. anatoliaca* farther east.

The nomenclature of the bees of Greece was hitherto rather vague. Usually the name "*cecropia*" was used as the only designation available, but the bees described in 1860 by Kiesenwetter undoubtedly orginated from the Hymettos, thus referring to a race from southern Greece.

All 13 samples from Bulgaria which were analyzed belong to *macedonica*, showing a relatively small variation. In Romania, Foti et al. (1965) describe a second autochthonous strain (besides the Carpathian), the "steppe bee", occurring in the plains of the lower Danube river and east of the Carpathian Mountains (province Moldavia). The morphometric data published by these authors correspond well with *macedonica*. Four samples of the Oberursel collection from Constanza and Tulcea were analyzed with the standard method: the coordinates of all three factors used fit exactly in the *macedonica* cluster. Six samples of "Ukrainian bees", derived from queens kindly provided to the Institute in Oberursel by the Research Station Rybnoe (USSR) were also located within the *macedonica* cluster.

Thus *A. m. macedonica* extends across the borders of five countries. Political frontiers have no significance, with one odd exception: in European Turkey typical *anatoliaca* is found with quite different characteristics. In Bulgaria, close to the Turkish border (Svielograd, Assenovgrad, Plovdiv) and in Greek Thracia, unhybridized macedonians are found. An explanation for this observation is very hard to give.

The morphometric data of *macedonica* differ in many characters, mostly only slightly, when compared with neighboring *carnica* bees from Yugoslavia (Table 14.2): smaller body and wing, but relatively longer legs; longer proboscis; broad metatarsi (oriental influence!); more slender than any other European race (St6 I = 86.7); shorter hair, slightly narrower tomenta; CI higher, and numerous (8 out of 11) slight differences in wing venation. *Macedonica* is also a dark bee but with more yellow at tergites and scutellum. The drones (two samples from the Black Sea coast in Romania) are also smaller than *carnica* drones and the CI is higher ($\bar{x}$ = 2.15 vs. 1.93). The behavior of *macedonica* (originating from Thessaloniki and Chalchidike) was described by Br. Adam: very gentle, not inclined to swarm, in spite of great colony strength, brood reduction in late summer, strong winter population, watery capping of honey cells, ample use of propolis. Several of these traits are in clear contrast to carniolans, see p. 257). Good wintering results are reported from England, but in Germany and Austria the temporary enthusiasm for Greek macedonians quickly disappeared after losses due to nosematosis. The Romanian "steppe bee" is described also as very gentle and little inclined to swarm. The number of swarm cells, however, is greater than in carpathians (76–195 queen cells; Foti et al. 1965). As in Greece, brood development was found to be 10 days later than in carniolans.

Greece as a whole, and especially Macedonia, is a country of prospering apiculture. Main honey source, besides the native flora and planted trees as *Robinia* and *Castanea*, is honeydew from a coccid (*Marschallina hellenica*) living on pines. On the small island of Thasos the unbelievable number of 60,000 colonies is concentrated in the narrow coastal strip (10 km long and 5 km deep) in late summer. On the peninsula Chalchidike 125,000 permanent colonies were found (39 colonies per km^2), representing perhaps the highest bee density of the world. For many inhabitants beekeeping is the only way to earn a living (Topalidis 1956). Migrations in Greece are practised mainly regionally: from west to east Macedonia or from Attica to the Peloponnes. This explains why *macedonica* and *cecropia* are still well separated (Ifantidis 1979).

14.2.5 *Apis mellifera carnica* Pollmann (1879):45 "Carniolan Bee"

In the valleys north and south of the Karawanken mountain chain, on both sides of the Austrian and Yugoslavian border, traditional beekeeping methods are found which differ greatly from those of Central and Northern Europe based on skep and vertical log hives.

The hives used in this region are horizontal boxes about 70 cm in length and 15×30 cm in diameter, opened from the front. These boxes are stacked in great piles of 30-100 hives. In this way the colonies keep each other warm during the cold winters, build up quickly in spring and are made to swarm early and frequently. This tendency is also enhanced by the small volume of the box.

Since this "Carinthian (Krainski) farmer's hive" was easily transported, migration up and down the mountain slopes was encouraged. The first report about long-distance migration with carniolan colonies is dated as far back as 1765, when the later famous bee biologist and apiculturist Anton Janjca traveled from his home in Slovenia to Vienna, taking along his 40 beehives by horse carriage. In the second part of the 19th century the railroad was used for transportation and bee dealers settled in this region, so that whoever needed swarms of bees in any part of Europe was able to order "Carinthian farmer's boxes" which were shipped by the thousands every year. Therefore, the Carniolan bee was soon well known all over the continent (Pollmann 1889).

The scientific name, derived from the disrict of Carnia, South of the Austrian-Italian border and from a mountain range in the same area ("Carniolan Alps"), was given to this bee by A. Pollmann in 1879.[1] Only 50 years later it became evident that this region of characteristical "Carniolan beekeeping" is only the westernmost corner of a much larger area of distribution of this race.

Distribution. From the visual aspect as well as from behavior and biometric analysis, the bees of northern SE Europe show the characteristics of *A. m. carnica*: the valley of the Danube from Vienna to the Carpathians, Austria south of the Alps and all of Yugoslavia including the coast of Dalmatia (Fig. 14.1). This was the identical result of earlier investigations by Goetze (1930, 1964) and Ruttner (1965), as well as of the more accurate multivariate analysis described later. To the NE and E the bees of Slovakia, South Poland and the Carpathian mountains clearly show the characteristics of *A. m. carnica*.

To the west, *A. m. carnica* borders on the area of *A. m. ligustica*, to the NW and N on that of *A. m. mellifera* and to the S and SE on that of *A. m. macedonica*. In the NW, the main chain of the Alps forms a clear borderline between *A. m. carnica* and *A. m. mellifera*, from E Tyrol to the hills W of Vienna (Ruttner 1950). This demon-

---

[1] *A. m. carnica* was first named by Pollmann in the 1st edition of his book, which is hard to find; 2nd edition 1889.

strates that even hills of moderate altitude can be an isolating factor for bee races. During the last few decades this borderline has disappeared more and more because carniolans are replacing the dark N European bee. In the east (CSSR, S Poland, Ukraine) a gradual transition is found to *A. m. mellifera*. All the selected *carnica* strains distributed during recent years, however, originated from the historical "*carnica*" region in SE Austria and in Slovenia, and to a lesser extent from Romania.

A. m. carnica is easily described (Fig. 14.3, Table 14.1): together with other Balkan races, *A. m. mellifera* and *caucasica* one of the largest races; dark body color; broad and dense tomenta; short cover hair; high CI and other characteristics of wing venation. Wing venation provides very clear discrimination from *caucasica* which is frequently subject of a mix-up on account of similarity in size, color and hair (see Figs. 14.3, 11.6). For discrimination from closely related races on the Balkan a very detailed analysis is needed: *cecropia* has an exceptionally long proboscis and high CI, *macedonica* is more slender with lower CI (Table 14.1). In the overall morphometric structure of this species, *A. m. carnica* occupies the major part in the C-branch of the "Y" (Fig. 10.7). Drones are large and dark, with grey hair on the thorax and a CI between 1.80 to 2.00. (Table 11.4)

The *carnica* area as a whole comprises a multitude of orographic positions and climatic conditions, sites on the coast as well as in 1600 m altitude, in dry, steppe-like plains and in humid forests. One factor, however, is predominant: a clear continental type of climate, with long, cold winters, quickly changing into hot, dry summers (Ruttner 1954b).

Local variation within *Apis mellifera carnica*. The next step after defining *A. m. carnica* morphometrically and geographically, is the examination of homogenity within the race. A number of varieties have been described in the *carnica* area, located mainly in marginal populations.

1. Besides the original Carniolan bee in Carinthia and northern Slovenia a "Lower Austrian bee", from a region east of Vienna was described (Berlepsch 1873), later known as "Sklenar strain". It was characterized by low swarming tendency and frequent leather-brown stripes on tergite 2.
2. The "Banat bee", was described by Grozdanic, (see Goetze 1964) from the plains near Timisoara in the three-country corner Romania – Yugoslavia – Hungary. *"Banatica"* shows a higher frequency of bright yellow bands (similar to Italians), but otherwise resembling a typical *carnica* exactly. Br. Adam (1983), who studied this bee in his apiary, did not detect any special trait in this strain.
3. The Dalmatian bee ("krasce pcela" of Mostar and the Adriatic coast; Table 14.2) shows narrow tomenta and higher irritability (Goetze 1964).
4. The "Carpathian bee" was extensively described and recorded as a geographic race of its own by Foti et al. (1963) and Barac (1977). This bee was selected and largely distributed among beekeepers in many countries especially in the USSR (Awetisjan et al. 1969) and therefore is well known in the apicultural world. Morphologically this population is characterized by somewhat shorter wings and hind legs and a higher CI when compared with the alpine *carnica*, which is almost identical in other measurements (Table 14.2, Foti et al. 1963; Awetisjan et al. 1969). These authors stress the gentle temperament and low swarming ten-

Fig. 14.3 *Carnica* worker bees. Photo H. Maag

Table 14.2 Morphometric characteristics of *carnica* sub-populations

	Alpine (Austria) 21	Pannonian (Hungary) 16	Carpathian (Rumania) 8	Dalmatia (Yougosl) 6	Ion. Islands (Greece) 5
n					
Proboscis	6.396	6.427	6.394	6.560	6.572
sd	0.154	0.142	0.147	0.146	0.102
Hind leg	8.102	8.069	8.072	7.954	8.104
sd	0.166	0.163	0.166	0.182	0.170
Terg 3+4	4.514	4.563	4.581	4.379	4.434
sd	0.123	0.115	0.117	0.134	0.105
Fore wing	9.403	9.265	9.225	9.177	9.176
sd	0.149	0.160	0.169	0.151	0.168
Cub index	2.589	2.776	2.790	2.508	2.859
sd	0.418	0.403	0.402	0.350	0.420
Stern 6 I	83.46	84.17	84.14	85.60	86.83
sd	2.91	3.13	3.37	3.48	3.52
Angle E9	23.12	23.43	23.65	22.75	22.85
sd	2.08	1.97	2.27	1.88	1.59
Angle G18	93.10	91.79	91.44	93.41	91.90
sd	2.77	2.72	3.07	2.51	2.46

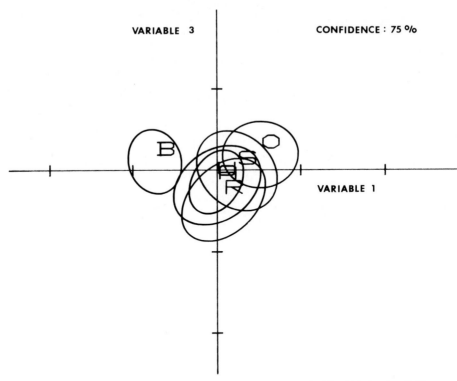

Fig. 14.4 DA "*carnica* within". Ellipses of confidence 75%. *A* Serbia, *B* Bulgaria, *H* Hungary, *O* Austria, *R* Romania, *S* Slovenia

dency of this bee. In spring, the development of the colonies is reported to be
10 days earlier than the colonies of the eastern Romanian steppe. Other *carnica*
strains are known as well, e. g., the "Troiseck" strain from Styria, Austria (Ruttner 1952c) and a local strain of the isolated highland of Pestar in Serbia (Vlatkovic 1956). The next step, therefore, had to be to study the intra-race variability
of carniolans. Ninety four samples were used for analyses from all *carnica* countries between the Alps and the Adriatic and the Black Sea, including Bulgaria,
but excluding Greece. Countries or regions were used as group determinants.
The results are shown in Figs. 14.4 and 14.5. Stepwise discriminant analysis gives
a barely asssorted picture puzzle with only two more or less recognizable clusters, Bulgaria (B) to the left and Austria (O) to the right. The calculation of
ellipses of confidence (75%) shows the same: a well separated group "Bulgaria",
an almost complete overlapping of the clusters in the center ("Hungary",
"Serbia") and a cluster "Austria" partly intersected by "Slovenia" and "Romania"(Fig. 14.4). Therefore, the samples were re-grouped at a larger scale according to zoogeographic zones (Fig. 14.5; Table 14.2):

1. Alpine (Austria and northern mountainous Slovenia)
2. Pannonic (Hungary, Romania to the Carpathians, most of Yugoslavia)
3. Pontic (Bulgaria)

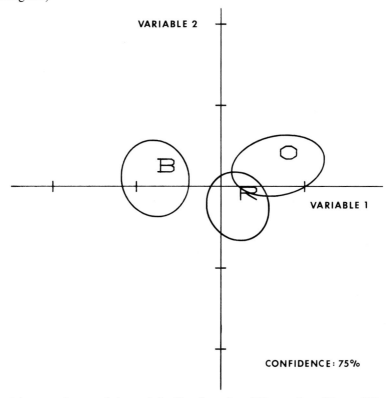

Fig. 14.5 DA of three carnica populations of the Danube valley. Ellipses of confidence 75%.
B Bulgaria (*A.m. macedonica*); *R* Hungary, western Romania including the Carpathians, Serbia:
pannonian population; *O* Austria and Slovenia: alpine population of *A. m. carnica*

This classification gives three separate groups with slight overlapping O/R. Since the cluster "Bulgaria" remains distinctly separated from the others in all analyses performed, but is completely intersected with the cluster "North Greece", these two groups are classified as the geographic race *A. m. macedonica* (p. 249). This race also includes the Romanian "steppe bee" (Foti et al. 1965) from the plains west of the Black Sea, the Ukrainian bee, as well as the bees from Yugoslavian Macedonia (Valandovo).

All other populations of the region show indications of diversification too vague to justify splitting into different races. Only two subunits are recognizable, although overlapping at 75% confidence: alpine *carnica* and pannonic *carnica* (including the Carpathian bee). *A. m. carnica* is a morphologically rather uniform bee, essentially limited to an area between the Alps, the Carpathians and the Dinarids. This uniformity may be explained by the recent colonization of this area after the last glaciation: climatic conditions excluded any existence of honeybees earlier than the Atlanticum, that is about 8000 years ago. This time span may be too short to evolve substantial morphological differences.

In Austria, the borderline *carnica-mellifera* was studied in detail, by examining 169 colonies with 14,000 worker bees and 5900 drones (Fig. 14.6; Ruttner 1969). No regional gradation of characters was found, in spite of great ecological differ-

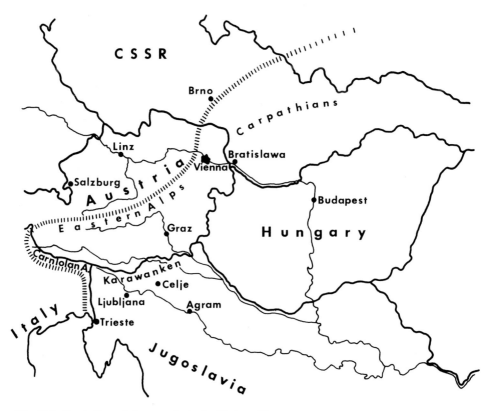

Fig. 14.6 Borderline of *A. m. carnica* in the Eastern Alps

ences. The selected strain "Troiseck" did not differ morphometrically from the non-selected population of the country.

A segregation more profound than in external characteristics is found in biological traits. Strains from the plains in Austria and Hungary differ from alpine strains in the annual cycle, swarming tendency, and wintering ability, as shown by repeated tests (Ruttner 1957; Goetze 1964, pers. unpubl. data). It could well be that the strain of the Dalmatian coast shows the Mediterranean type of brood rhythm. These biological characteristics of local ecotypes are of special importance to apiculture.

The annual cycle of the *carnica* colonies is well adapted to climatic conditions (Ruttner 1954b). This bee survives long and cold winters with a relatively small winter cluster. In most regions and years the broodless period lasts from the end of September to the end of February-March. Spring build-up occurs very rapidly, making the *carnica* a "spring honey producer par excellence" (Fig. 10.4; Br. Adam 1983). During the summer, *carnica* adapts brood production sensibly to the availability of nectar and pollen.

As far as defensive behavior is concerned, *carnica* can be considered one of the most gentle races. This bee remains quietly on the comb when disturbed; very little propolis is used.

Swarming tendency may be considerable when compared with related races, like *A.m. ligustica*, although this trait (as all others described) shows great variability and is easily influenced by selection (Ruttner 1982). Impressive is the capacity for homing: carniolan field bees have no problem finding the entrance to their hive even if the colonies are arranged close together. In an experiment with eight *carnica* and eight *mellifera* nuclei stacked in four superposed rows in alternative arrangement, drifting from *mellifera* to *carnica* colonies was more than four times as much as vice versa (13.7 vs. 2.9%; Ruttner 1954a). The whole search strategy is focused on the position of an object rather than on color (Lauer and Lindauer 1973). The dance rhythm during communication is different from, e.g., *A.m. ligustica*: waggle dance starts at a greater distance (100 m) and the number of waggle runs per time unit decreases more slowly with increasing distance (Fig. 4.9), indicating frequent long-distance flights. Thus *A.m. carnica*, being especially well investigated, is different from comparable races in quite a number of behavioral characteristics.

Though the original *carnica* area extends no farther north than 50° N lat., this race (together with *A.m. mellifera*) is one with the best wintering abilities, as shown by many comparative tests. This may be explained by the long and very cold winters (continental type) in the Alps, the Balkan, and the Carpathian mountains. Even in the lowlands of the Danube valley temperatures below $-30°C$ are not rare (Ruttner 1954b). Therefore, *carnica* strains have been well established in Norway, in the USSR, and in Canada. Since the middle of the 1930's, the bees of Central Europe, which originally belonged to the "European Brown bee", *A.m. mellifera*, are being continually replaced by *carnica* strains from Austria and Yugoslavia.

References

Abushady AZ (1919) The faults of the Punics. Bee World 1: 77–78

Abushady AZ (1949) Races of bees. In: Dadant, The Hive and the Honey Bee: 11–20 Dadant Sons, Hamilton Illinois

Adam Brother (1983) In search of the best strains of bees. Dadant Sons, Hamilton Illinois

Adlakha RL (1971) Peculiarities of mating of honeybee queens of *Apis mellifera* and *Apis cerana* and beginning of oviposition. Proc Int Beekeep Congr 23: 434–439

Adlakha RL, Sharma OP (1971) Introducing honeybee queens of *Apis mellifera* into colonies of *Apis indica.* Proc Int Beekeep Congr 23: 415

Adlakha RL, Sharma OP (1974) Interspecific robbing. Am Bee J 114: 248–249

Agwu COC, Akanbi TO (1985) A palynological study of honey from four vegetation zones of Nigeria. Pollen Spores 27: 335–348

Ahmad R (1982) Introduction of *Apis mellifera* L and factors affecting its establishment in Pakistan. Rep Pak Agric Res Comm

Ahmad R (1984) Country status report on beekeeping in Pakistan. Proc Expert Consult Bangkok: 203–210. FAO Rome

Ahmad R, Muzzafar N, Ali Q (1985) Biological control of the wax moths *Achroia grisella* F and *Galleria mellonella* L by propagation of the parasite *Aplateles galleriae* Wilk. (Hym., Braconidae) in Pakistan. Apiacta 18: 15–20

Akahira J, Sakagami SF (1958) Zum gegenwärtigen Zuchtzustand der japanischen Honigbiene in Kyushu, Süd-Japan. Studien zur japanischen Honigbiene, *A. indica cerana* Fabr. II. Z Bienenforsch 4: 87–96

Akahira F, Sakagami SF (1959) Notes on the difference in some external characteristics between Japanese and European honeybees. Studies on the Japanese honeybee *A. cerana cerana* F III. Annot Zool Jpn 32: 35–42

Akratanakul P (1976) Honeybees in Thailand. Am Bee J 116: 121–126

Akratanakul P (1977) The natural history of the dwarf honeybee, *Apis florea* F, in Thailand. PhD Thesis, Cornell Univ Ithaca NY

Akratanakul P (1984) Beekeeping industry with *Apis mellifera* in Thailand. Proc Expert Consult Bangkok: 222–232. FAO Rome

Alber M (1956) La misura dele celle caratteristica razziale. Apic Ital 23: 35–42

Alpatov WW (1925) Über die Verkleinerung der Rüssellänge der Honigbiene von Süden nach Norden hin. Zool Anz 65: 103–111

Alpatov WW (1929) Biometrical studies on variation and races of the honeybee *Apis mellifera* L. Rev Biol 4: 1–57

Alpatov WW (1932) Some data on the comparative biology of different bee races. Bee World 13: 138–139

Alpatov WW (1933) South African bees biometrically investigated. Bee World 14: 62–64

Alpatov WW (1935a) Contribution to the study of the variation in the honeybee, IV. The Egyptian honeybee and its position among other bees. Bull Soc Nat Mosc S Biol 44: 284–291 (in Russian)

Alpatov WW (1935b) The cubital cells on the wings of different forms of the genus *Apis.* Zool J 14: 660–671 (Russian with English summary)

Alpatov WW (1938a) Contribution to the study of variation in the honeybee, V. Number of egg tubules in ovaries of worker bees in the genus *Apis* as related to the evolution of polymorphism. Zool J 17: 241–245

Alpatov WW (1938b) Contribution to the study of variation in the honeybee, VI. Carniolan and Crimean bees and their places among other forms of *Apis mellifera.* Zool.J 17: 473-481 (Russian with English summary)

Alpatov WW (1946) Mutual assistance of insects and entomophile plants as example of an interspecific symbiosis. Zool. J 25: 325-328 (Russian)

Alpatov WW (1948) The races of honeybees and their use in agriculture. Sredi Prirodi 4, 1-65 (Russian)

Alpatov WW (1976) A fatal mistake in the determination of a race of honeybees. Priroda 5: 72-73

Anderson RH (1963) The laying worker in the Cape honeybee, *Apis mellifera capensis.* J Apic Res 2: 85-92

Anderson RH (1976) Some aspects of the Cape bee biology. Proc Int Apimondia Symp Afr Bees 141-151

Anderson RH (1981) Queens and queen rearing. S Afr Bee J 53: 3-12

Anderson RH, Buys B, Johannsmeier MF (1973) Beekeeping in South Africa. Dept Agri tech Bull 394. Pretoria

Andguladze DI (1971) History, success and perspectives of apiculture in the socialist sector of the Grusinian SSR. Proc Int Beekeep Congr 23: 304-308

Ankinovich GB (1957) An experiment on acclimatizing Indian bees and their use for pollinating agricultural crops. Dokl TSKhA 30: 307-310

Armbruster L (1938) Versteinerte Honigbienen aus dem obermiocänen Randecker Maar. Arch Bienenkd 19: 1-48, 97-133

Ashmead WH (1904) Remarks on honeybees. Proc ent soc Washington 6: 120-123

Atwal AS (1968) The introduction of *Apis mellifera* queens into *Apis indica* colonies and the associated behaviour of the two species. Indian Bee J 30: 41-56

Atwal AS (1970) Insect pollinators of crops. Punjab Agric Univ Press, Ludhiana

Atwal AS, Dhaliwal GS (1969) Some behavioural characteristics of *Apis indica* F and *Apis mellifera* L. Indian Bee J 31: 83-90

Atwal AS, Sharma OP (1970a) Studies on the performance of five strains of *Apis mellifera* as compared with *Apis indica* under Nagrora conditions. J Apic Res 7: 477-486

Atwal AS, Sharma OP (1970b) Brood rearing activity of Italian honeybees *A. mellifera* in the Punjab plains at Ludhiana. Indian Bee J 32: 62-67

Atwal AS, Sharma OP (1971) The dominance of *Apis mellifera* over *Apis indica.* Am Bee J 111: 343

Awetisjan GA (1978) Apiculture. Apimondia Publishing House, Bucharest

Awetisjan GA, Maximenko IW (1979) Interline variability of honeybee's resistance against Nosema. Proc Int Beekeep Congr 27: 258-261

Avetisjan GA, Gubin WA, Davydenko IK (1969) Selection of Carpathian bees. Proc Int Beekeep Congr 22: 366-371

Badino A, Celebrano G, Manino A (1984) Population genetics of Italian honeybee *Apis mellifera ligustica* Spin and its relationship with neighbouring subspecies. Bull Mus Sci Nat Torino 2: 571-584

Badino A, Celebrano G, Manino A, Longo S (1985) Enzyme polymorphism in the Sicilian honeybee. Experientia 41, 752-754

Bährmann R (1961a) Über den Bau des Begattungsschlauches der vier *Apis*-Arten. Leipziger Bienenztg 75: 18-20

Bährmann R (1961b) Vergleichend-morphologische Untersuchungen an *Apis cerana* und *Apis mellifica.* In Oschmann, Eine bienenkundliche Reise in die VR China. Arch Geflügelz Kleintierkd 10: 244-257

Bährmann R (1965) Merkmalsstudien und infraspezifische Verwandtschaftsverhältnisse bei *Apis mellifica* L. Z Bienenforsch 8: 17-47

Bährmann R (1967) Vergleichend-morphologische Untersuchungen an Arbeitsbienen und Drohnen von *Apis cerana* F. Z Bienenforsch 9: 32-46

Baldensperger PhJ (1922) Sur l'apiculture en orient. Proc Congr Int d'Apic 6: 59-64

Baldensperger PhJ (1924) North African bees II. Bee World 5: 189-190

Baldensperger PhJ (1928) Bees in their natural state and with beekeepers. Bee World 9: 173-174

Baldensperger PhJ (1932) Variétés d'abeilles en Afrique du Nord. Proc Congr Int Entomol 5: 829-839

Baltzer (1956) in K.v.Frisch 1967

Barac I (1977) Preserving the genetic pool of *Apis mellifera carpatica*. Proc Int Beekeep Congr 26: 270–274

Barbier E (1969) Present situation and development program of apiculture in Morocco. Proc Int Beekeep Congr 22: 372–374

Bartholomew GA (1982) A matter of size: an examination of endothermy in insects and terrestrial vertebrates. In B. Heinrich (ed) Insect thermoregulation 46–78

Battesti MJ (1980) Etude biométrique de colonies d'abeilles corses. Mémoire de D. E. A. d'écologie Méditerranéenne, Univ Marseille

Ben-Neriah A (1958) Result of the introduction of the Italian bee during the first 10 years of Israel's statehood. Proc Int Beekeep Congr 17: 262–264

Berlepsch A (1873) Die Biene und ihre Zucht auf beweglichen Waben. 3rd ed J. Schneider, Mannheim

Bilash GD (1967) Organisation and working methods in the distribution of the honeybee races in the USSR. Proc Int Beekeep Congr 21: 387–390

Bilash GD, Makarov II, Sedich AW (1976) Geographic classification of honeybee races in the USSR. Apimondia Symp Genetics Selection Reproduction: 140–150

Blandford EJ (1923) Chinese bees as we find them. Bee World 5: 104–106

Blum MS (1986) Specificity in chemisocial signals: evolution of fine tuning mechanism. Proc INT IUSSI Congr 10 (in press)

Blum R (1956) The bees in the Near East. Proc Int Beekeep Congr 16: 28

Boch R (1957) Rassenmäßige Unterschiede bei den Tänzen der Honigbiene. Z vgl Physiol 40: 289–320

Bodenheimer FS (1941) Studies on the honeybee and beekeeping in Turkey. Merkez Ziraat Macadela Enstitüsü, Ankara

Bornus L, Gromisz M, Novakovski I (1976) Utilization of morphological characters in the taxonomy of honeybees. Proc Apimondia Symp Genetics Selection Reproduction: 214–216

Böttcher FK (ed) (1984) Krankheiten der Biene. Ulmer, Stuttgart

Böttcher FK, Mautz D, Weiss K (1975) Die Tätigkeit der Bayerischen Landesanstalt für Bienenzucht Erlangen im Jahre 1974. Imkerfreund 60: 76–77

Brandes C (1984) Selektion des Lernverhaltens bei der Kap-Honigbiene *Apis mellifera capensis* Esch. Apidologie 15: 273–274

Brandes C, Moritz RFA (1983) Evaluation of the heritability of learning behavior in *A. mellifera* L using the proboscis extension reflex. Verh Dtsch Zool Ges 1983: 258B

Breadbear N (1986) *Apis florea* in Africa. Newsl Beekeeping Trop Countries 8: 1. IBRA Cardiff

Brice WC (ed) (1978) The environmental history of the Near and Middle East since the last Ice age. Academic Press, London New York

Bronard J, "Campèche" (1973). L'introduction des abeilles aux Mascareignes. L'Express (Paris) July 20, 1973

Brown RH (1978) Adansonii management. Proc Apimondia Symp Apic in hot climate 187–192

Buchner R (1953) Beeinflussung der Größe der Arbeitsbiene durch Raum- und Nahrungsmangel während der Larvenzeit. Roux's Arch Entw Mech 146: 544–579

Buco SM, Rinderer T, Sylvester A, Collins A, Lancaster V, Crewe R (1987) Morphometric differences between Africanized and South African (*Apis mellifera scutellata*) honeybees. Apidologie (in press)

Büll R (1961) Vom Wachs. Band I/5. Farbwerke Hoechst AG

Burgett DM, Krantz GW 1984) The future of the European honeybee (*Apis mellifera*) in Southeast Asia: constants of parasitism. Proc Expert Consult Bangkok: 34–40. FAO, Rome

Burleigh R, Whalley P (1983) On the relative geological age of amber and copal. J Nat Hist 17: 919–921

Butani K (1950) An *Apis dorsata* colony in New Delhi. Indian Bee J 12: 115

Butler CG (1954) The world of the honeybee. Collins London

Butler CG, Calam DH, Callow RK (1967) Attraction of *Apis mellifera* drones by the odors of the queens of the two other species of honeybees, *Apis cerana* ssp. *indica* and *Apis florea*. Nature (Lond) 213: 423–424

Buttel-Reepen H (1903)Die stammesgeschichtliche Entstehung des Bienenstaates. Thieme, Leipzig

Buttel-Reepen H (1906) Apistica. Beiträge zur Systematik, Biologie, sowie zur geschichtlichen und geographischen Verbreitung der Honigbiene (*Apis mellifica* L), ihrer Varietäten und der übrigen *Apis*-Arten. Veröff Zool Mus Berlin 118–120

Buttel-Reepen H (1907) Psychobiologische und biologische Beobachtungen an Ameisen, Bienen und Wespen. Naturwiss Wochenschr NF 6: 465–478

Buttel-Reepen H (1921) Zur Lebensweise der ägyptischen Biene (*Apis mellifica fasciata* Latr.) sowie einiges zur Geschichte der Bienenzucht. Arch Bienenkd 3: 19–67

Butzer KW (1978) The late prehistoric environmental history of the Near East, In: Brice WC (ed) The environmental history of the Near and Middle East since the last Ice age. Academic Press, London New York pp 5–12

Cadapan EP (1984) Beekeeping with *Apis mellifera* in the Philippines. Proc Expert Consult Bangkok: 211–216. FAO Rome

Chandler MT (1976)The African honeybee, *Apis mellifera adansonii*: the biological basis of its management. Proc Conf Apic Trop Climate 61–68. IBRA, Gerrards Cross

Chonduri MM (1940) Beekeeping in Orissa. Indian Bee J (cited from Bee World 1941: 66)

Church LJ, White R (1980) Beekeeping and honey production in Cyprus. Proc Int Conf Apic Trop Climate 2: 110–118. Agric Res Inst New Delhi

Clauss B (1983) Bees and beekeeping in Botswana. Report Min Agric. Botswana

Cochlov BP (1916) Investigations on the length of the bee tongue. Min of Agric Petrograd 17–41 (Russian)

Cockerell TDA (1906) New Rocky Mountain bees, and other notes. Canad Entomol 38: 160–166

Cockerell TDA (1907) A fossil honeybee. Entomologist (Lond) 40: 227–229

Cockerell TDA (1908) Descriptions and records of bees. 20 Ann Mag Nat Hist Lond 2: 323–334

Cornuet JM (1982a) Représentation graphique de populations multinormales par des ellipses de confiance. Apidologie 13: 15–20

Cornuet JM (1982b) The MDH polymorphism in some westmediterranean honeybee populations. Int IUSSI Congr Boulder

Cornuet JM (1983) Reproduction génétique et sélection de l'abeille. Bull Tech Apic 1983, Acad d'Agric France

Cornuet JM, Fresnaye J, Tassencourt L (1975) Discrimination et classification de populations d'abeilles à partir de caractères biométriques. Apidologie 6: 145–187

Cornuet JM, Fresnaye J, Lavie P (1978) Etude biométrique de deux populations cévenoles. Apidologie 9: 41–55

Cornuet JM, Albisetti J, Mallet N, Fresnaye J (1982) Etude biométrique de deux populations landaises. Apidologie 13: 3–13

Crewe RM (1984) Differences in behaviour and morphology between *capensis* and *adansonii*. S Afr Bee J 56 (1): 16–20

Daly HV (1985) Insect morphometrics. Ann Rev Entomol 30: 415–438

Daly HV, Balling St (1978) Identification of Africanized honeybees in the western hemisphere by discriminant analysis. J Kansas Entomol Soc 51: 857–869

Daly HV, Hoelmer K, Norman P, Allen T (1982) Computer-assisted measurement and identification of honeybees (*Hymenoptera: Apidae*). Ann Entomol Soc Amer 75: 591–594

Danilenko PI (1975) Apiculture in the Far East and its further development. Proc Int Beekeep Congr 25: 180–183

Danilova LV (1960) Some morphological pecularities of the Far Eastern wild bees. Pchelovodstvo 37 (11): 26–27 (Russian)

De Kok CV (1976) Beekeeping in the Letaba district. Proc Int Apimondia Symp Afr bees: 91–95

Delfinado MD (1963) Mites of the honeybee in South-East Asia. J Apic Res 2: 113–114

Delfinado-Baker MD (1982) New records for *Tropilaelaps clareae* from colonies of *Apis cerana indica*. Am Bee J 122: 382

Delfinado MD, Baker EW (1974) Varroaidae, a new family of mites in honeybees. Wash Acad Sci 64: 4–10

Delfinado-Baker MD, Underwood BA, Baker EW (1985) The occurrence of *Tropilaelaps* mites in brood nests of *Apis dorsata* and *Apis laboriosa* in Nepal, with description of the nymphal stages. Am Bee J 125: 703–706

Deodikar GB, Thakar CV (1966) Cyto-genetics of Indian honeybee and bearing on taxonomic and breeding problems. Indian J Genet 36 A: 386–393

Deodikar GB, Thakar CV, Tonapi KV (1958) Evolution in the genus *Apis* and its bearing on breeding better strains of Indian bees. Proc Int Beekeep Congr 17: 245–250

Dhaliwal GS, Atwal AS (1970) Interspecific relations between *Apis cerana indica* and *Apis mellifera*. J Apic Res 9: 53–59

Dhaliwal GS, Sharma PL (1973) The foraging range of the Indian honeybee on two crops. J Apic Res 12: 131–134

Dhaliwal GS, Sharma PL (1974) The foraging range of the Indian honeybee. J Apic Res 13: 137–141

Dinabandhov CL, Dogra GS (1980) The acarine induced decline in Indian honeybee *Apis cerana indica* F and its control in Kulu Valley. Int Conf Apic Trop Climate 2: 434–441. New Delhi

Divan VV, Salvi SR (1965) Some interesting behavioural features of *Apis dorsata* F. Indian Bee J 27: 52

Douglas JC (1886) The hive-bee indigenous to India and the introduction of the Italian bee. J Asiat Soc Bengal II Nat Sci 55: 83–96

Douhet M (1965) Beekeeping in Madagascar. Proc Int Beekeep Congr 20: 690–710

Drescher W (1975) Bienennutzung in Tansania. Allg Dtsch Imkerzeitg 9: 117–122

Drory E (1888) Aus meinem Tagebuch. Apistische Notizen während einer Reise um die Erde. (cit Buttel-Reepen 1903)

Dubois L, Collart E (1950) L'apiculture au Congo Belge et au Ruanda-Urundi. Min Col Dir Agric, de l'élevage et de la colonisation. Bruxelles

DuPraw E (1964) Non-Linnean taxonomy. Nature (Lond) 202: 849–852

DuPraw E (1965a) Non Linnean taxonomy and the systematics of honeybees. Syst Zool 14: 1–24

DuPraw E (1965b) The recognition and handling of honeybee specimens in non-Linnean taxonomy. J Apic Res 4: 71–84

Dutton RH, Free JB (1979) The present status of beekeeping in Oman. Bee World 60: 176–185

Dutton RH, Ruttner F, Berkeley A, Manley MJD (1981) Observations on the morphology, relationship and ecology of *Apis mellifera* of Oman. J Apic Res 20: 201–214

Dyer FC (1985a) Mechanisms of dance orientation in the Asian honeybee *Apis florea*. J Comp Physiol A: 183–198

Dyer FC (1985b) Nocturnal orientation by the Asia honeybee, *Apis dorsata*. Anim Behav 33: 769–774

Dyer FC, Seeley TD (1987) Interspecific comparisons of endothermy in honeybees *(Apis)*: deviations from the expected size-related patterns. J Exp Biol (in press)

El Banby MA (1963a) Biological investigations of the Egyptian honeybee race. Activities of brood rearing. Proc Int Beekeep Congr 19: 31–33

El Banby MA (1963b) Biometrical investigations of the Egyptian bee. Proc Int Beekeep Congr 19: 33–34

El Banby MA (1977) Biometrical studies on the local honeybee of the Libyan Arab People's Socialist Jamachiriya. Proc Int Beekeep Congr 26: 269

Eldredge N, Gould SJ (1972) Punctuated equilibria: an alternative to phyletic gradualism. in: Schopf TJM (ed) Models in paleobiology. Freeman, Cooper & Co, San Francisco pp 82–115

Enderlein G (1906) Neue Honigbienen und Beiträge zur Kenntnis der Verbreitung der Gattung *Apis*. Stettiner Entomol Zeitg 67: 331–344

Engels W (1973) Die Evolution der Haemolymph-Protein-Spektren als entwicklungsphysiologisches Merkmal bei den weiblichen Kasten der vier rezenten *Apis*-Arten *florea, dorsata, cerana* und *mellifera*. Proc Int Beekeep Congr 24: 327–332

Escholtz JF (1822) Entomographien. Vol 1. Reimer, Berlin

Eskov EK (1976) Time structure of acustic signals – a race specific characteristic of the honeybee. Proc Apimondia Symp Genetics Selection Reproduction 175–178

Fahrenhorst H (1977a) Nachweis übereinstimmender Chromosomen-Zahlen (n = 16) bei allen vier *Apis*-Arten. Apidologie 8: 89–100

Fahrenhorst H (1977b) Chromosome number in the tropical honeybee species *Apis dorsata* and *Apis florea*. J Apic Res 16: 56–58

Fang Tsung-Deh (1956) Migration, defense and ventilation of hives in Chinese bees. Proc Int Beekeep Congr 16: 38–39

Fang Yue-Zhen (1984) The present status and development of keeping European bees *(Apis mel-*

lifera) in tropical and subtropical regions of China. Proc Expert Consult Bangkok 142-147 FAO Rome

FAO (ed) (1984) Proceedings of the expert consultation on beekeeping with *Apis mellifera* in tropical and subtropical Asia. FAO Rome

Field OS (1980) Beekeeping and honey production in Yemen Arab Republic. Rep Field Honey Farms, Thame England

Firsow WS (1976) Variability of venom and rectal glands in honeybees. Proc Apimondia Symp Genetics Selection Reproduction 179-181

Fletcher DJC (1978a) The African bee, *Apis mellifera adansonii*, in Africa. Ann Rev Entomol 23: 151-171

Fletcher DJC (1978b) Management of *Apis mellifera adansonii* for honey production in southern Africa. Proc Apimondia Symp Apic Hot Climate 86-89

Fletcher DJC, Tribe GD (1977a) Natural emergency queen rearing by *Apis mellifera adansonii* and its relevance for successful queen production. Proc. Apimondia Symp Afric Bees 161-168

Fletcher DJC, Tribe GD (1977b) Swarming potential of the African bee, *Apis mellifera adansonii*. Proc. Apimondia Symp Afric Bees 25-37

Foti N, Lungu M, Pelimon P, Barac I, Copaitici M, Mirza E (1965) Studies on the morphological characteristics and biological traits of the bee populations in Romania. Proc Int Beekeep Congr 20: 182-188

Francis BR, Blanton WE, Nunamaker RA (1985) Extractable surface hydrocarbons of workers and drones of the genus *Apis.* J Apic Res 24: 13-26

Fraser HM (1951) Beekeeping in antiquity. Univ London Press, London

Free JB (1970) Insect pollination of crops. Academic Press, London New York

Free JB (1981) Biology and behaviour of the honeybee *Apis florea*, and its possibilities for bee-keeping. Bee World 62: 46-59

Free JB, Spencer-Booth Y (1961) Effect of temperature on *Apis indica* workers. Nature 190: 933

Free JB, Williams IH (1979) Communication by pheromones and other means in *Apis florea* colonies. J Apic Res 18: 16-25

Fresnaye J (1981) Biometrie de l'abeille, 2nd ed. OPIDA, Echauffour

Friese H (1909) Die Bienen Afrikas nach dem Stande unserer heutigen Kenntnisse. Jena Fischer: Jena

Frisch K (1947) Duftgelenkte Bienen im Dienste von Landwirtschaft und Imkerei. Springer Verlag, Berlin Heidelberg New York

Frisch K (1967) The dance language and orientation of bees. Harvard Univ Press, Cambridge MA

Gabriel B (1978) Klima- und Landschaftswandel der Sahara. In (ed) v Gagern, Sahara Ausstellungskatalog Köln 22-34

Gadbin C (1976) Apercu sur l'apiculture traditionelle dans le sud du Tchad. J agric trop botan appliqu 23: 101-115

Gadbin C (1980) Les plantes utilisées par les abeilles au Tchad méridionale. Apidologie 11: 217-254

Gadbin C, Cornuet JM, Fresnaye J (1979) Approche biométrique de la variété locale d'*Apis mellifica* L dans le sud tchadienne. Apidologie 10: 137-148

Gary NE (1975) Activities and behavior of honeybees. In Dadant & Sons (ed) The Hive and the Honeybee. Hamilton Illinois

Gasanov SO (1967) Flower migration and spezialisation of honeybees of various races. Proc Int Beekeep Congr 21: 281-284

Gautam SK (1984) Country status report on beekeeping in Nepal (discussion). Proc. Expert Consult Bangkok: 198-202. FAO Rome

Gaydar WA (1976) Use of package bees of Carpathian bees in Siberia. Proc Apimondia Symp Genetic Selection Reproduction: 274-276

Gebreyesus M (1976) Practical aspects of bee management in Ethiopia. Int Conf Apicult Trop Climate 1: 69-78. IBRA, Gerrards Cross

Genduso P, Alber M (1973) Ricerche sulle ape Siciliane (*Apis mellifera sicula* Grassi) I Produzione di celle reale. Boll Ist Entomol Agr Palermo 8: 45-48

Gerstäcker A (1862) Über die geographische Verbreitung und die Abänderungen der Honigbiene nebst Bemerkungen über die ausländischen Honigbienen der alten Welt. In: Buttel-Reepen 1906 Apistica

Glushkov MN (1954) Diminution of cell size from north to south. Pchelovodstvo 1954/1: 21 (Russian)

Goetze G (1930) Variabilitäts- und Züchtungsstudien an der Honigbiene mit besonderer Berücksichtigung der Langrüßligkeit. Arch Bienenkd 11: 135–274

Goetze G (1940) Die beste Biene. Liedloff Loth Michaelis Leipzig

Goetze G (1949) Imkerliche Züchtungspraxis. Landbuchverlag Hannover

Goetze G (1964) Die Honigbiene in natürlicher und künstlicher Zuchtauslese. Parey, Hamburg

Goncalves LS (1979) Directed selection of two inbred lines of *Apis mellifera* L. Proc Int Beekeep Congr 27: 335

Gontarski H (1935) Über die phaenotypische Beeinflußbarkeit der Arbeiterinnen von *Apis mellifera* durch die Brutzellengröße. Z Morph Ökol Tiere 29: 455–471

Gorbachev AN (1916) The grey mountain Caucasian bee (*Apis mellifera caucasica*) and its place among other bees. Tiflis (Russian with English summary)

Goug YF (1983) The natural beekeeping conditions and honeybee races in China. J Fujian Agric Coll 12: 241–249

Gough L (1928) Apistischer Brief aus Südafrika. Der Bienenvater 60: 30–32

Gould JD (1982) Why do honeybees have dialects? Behav Ecol Sociobiol 10: 53–56

Gould JD, Dyer FC, Towne WF (1985) Recent progress in the study of the dance language. Fortschr Zool 31, Exp Behav Ecol 141–161

Goyal NP (1974) *Apis cerana indica* and *Apis mellifera* as complimentary to each other for the development of apiculture. Bee World 55: 98–101

Goyal NP, Atwal AS (1971) Comparison of the way of communication with *Apis indica* F and *Apis mellifera* L. Proc Int Beekeep Congr 23: 341–342

Goyal NP, Atwal AS (1977) Wing beat frequencies of *Apis cerana indica* and *Apis mellifera*. J Apic Res 16: 47–48

Goyal NP, Atwal AS (1979) Introduction of exotic honeybees (*Apis mellifera*) in India. Indian Bee J 41/3–4: 39–45

Grassi B (1871) Saggio di una monografia delle Api d'Italia. Ann Agric Siciliana NS 3: 277–281 (Palermo)

Gregor H.-J (1982) Die jungtertiären Floren Süddeutschlands. Enke, Stuttgart

Grimaldi D, Underwood BA (1986) *Megabraula*, a new genus for two new species of Braulidae (*Diptera*), and a discussion of braulid evolution. Syst Entomol 11: 427–438

Gromisz M (1962) Season variation of wing measurements and the cubital index of honeybees. Pszczelnicze Zesz Nauk 6: 113–120

Grout RA (1937) The influence of size of brood cell upon the size and variability of the honeybee (*Apis mellifera* L). Res Bull Agr Exp Sta Iowa State Coll 218: 260–280

Grozdanic SS (1926) Die "gelbe" Banater Biene. Acta Soc Entomol Serb Beograd 1: 45–60 (Serbian with German summary)

Grünewald G, Höller E, Stranz D (1983) Länder und Klima: Afrika. Brockhaus, Wiesbaden

Guiglia D (1964) Missione 1962 del Prof. Giuseppe Scortecci nell'Arabia meridionale. Hymenoptera: Tiphiidae, Vespidae, Eumenidae, Pompilidae, Sphecidae, Apidae. Atti Soc ital Sci nat 103: 305–310

Guiglia D (1968) Missione 1965 del Prof. Giuseppe Scortecci nello Yemen (Arabia meridionale). Hymenoptera: Tiphiidae, Vespidae, Pompilidae, Sphecidae, Apidae. Atti Soc ital Sci nat 107: 159–167

Guy RD (1976) Commercial beekeeping with *Apis mellifera adansonii* in intermediate and moveable-frame hives. Apiculture in trop climate. IBRA Gerrards Cross

Haccour P (1960) Recherches sur la race d'abeille saharienne au Maroc. C Soc Sci Nat Physique Maroc 6: 96–98

Hadorn H (1948) Betrachtungen über wilde Bienen in Sumatra. Schweiz Bienenztg 1948: 309–314

Hagmeister A (1973) Immunologischer Vergleich der Haemolymphproteine der vier rezenten *Apis*-Arten *mellifera, cerana, florea* und *dorsata*. Proc Int IUSSI Congr 7: 151–153

Hänel H, Koeniger N (1986) Possible regulation of the reproduction of the honeybee mite *Varroa jacobsoni* (*Mesostigmata: Acari*) by a host's hormone: juvenile hormone III. J Insect Physiol 32: 791–798

Hänel H, Ruttner F (1985) The origin of the pore in the drone cell capping of *Apis cerana* Fabr. Apidologie 16: 157–164

Hanser G, Rembold H (1960) Über den Weiselzellenfuttersaft der Honigbiene IV. Jahreszeitliche Veränderungen im Biopteringehalt des Arbeiterinnenfuttersaftes. Hoppe Seyler's Z Physiol Chemie 319: 200-205

Harbo JR (1986) Effect of population size on brood production, worker survival and honey gain in colonies of honeybees. J Apic Res 25: 22-29

Harbo JR, Bolten AB, Rinderer TE, Collins EM (1981) Development periods for eggs of Africanized and European honeybees. J Apic Res 20: 156-159

Heinrich B (1985) The social physiology of temperature regulation in honeybees. In: (ed) Lindauer M Hölldobler B Exper Behav Ecol; Fortschr Zool 31: 393-406

Hemmling C, Koeniger N, Ruttner F (1979) Quantitative Bestimmung der 9-Oxodecensäure im Lebenszyklus der Kapbiene (*Apis mellifera capensis* Esch). Apidologie 10: 227-240

Hepburn HR (1986) Honeybees and wax: an experimental natural history. Springer-Verlag, Berlin Heidelberg New York Tokyo

Hicheri H, Bouderbala M (1969) Tunisian apiculture. Proc Int Beekeep Congr 22: 440-443

Hoefer I, Lindauer M (1975) Das Lernverhalten zweier Bienenrassen unter veränderten Orientierungsbedingungen. J Comp Physiol 99: 119-138

Horn EP (1975) Mechanisms of gravity processing by leg and abdominal gravity receptors in bees. J Insect Physiol 66: 663-679

Horne Ch, Smith F (1870) Notes on the habits of some hymenopterous insects of the NW provinces of India. Trans Zool Soc 7: 161-196

Hoshiba H, Okada J, Kusanagi A (1981) The diploid drone of *Apis cerana japonica* and its chromosomes. J Apic Res 20: 143-147

Hukusima S, Inoue A (1964) Respiration rates of Japanese and European honeybees during various stages of development. Kansey-Byochyngay-Kenkyakai Ho 6: 36-44

Husain SW (1938) How honey is extracted from the combs of the Giant bee of India. Bee World 19: 139-140

Huxley J (1938) Clines: an auxiliary taxonomic principle. Nature(Lond) 142: 219-220

Huxley J (1939) Clines: an auxiliary method in taxonomy. Bijdr Dierk 27: 491-518

Ifantidis M D (1979) Morphological characteristics of the Greek honeybee *Apis mellifica cecropia*. Proc Int Beekeep Congr 27: 285-291; 292-294

Inci A (1980) Entwicklung der Varroatose-Kontrolle in der Türkei. Proc Apimondia Symp Diagnosis and Therapy of Varroatosis: 87-91

Inoue A (1962) Preliminary report on rearing of Japanese honeybee queens in colonies of the European honeybee. Bee Science Nagoya 3: 10

Inoue A (1985) Relationship between *Apis cerana japonica* and *Apis mellifera*, with special reference to beekeeping technology. Abstr Int Beekeep Congr 30: 51-52

Jack RW (1916) Parthenogenesis amongst the worker of the Cape honeybee: Mr. G.W. Onions experiments. Trans Entomol Soc London 1916/17: 396-403

Jagannadham B, Goyal NP (1980) Morphological and behavioural characteristics of honeybee workers reared in combs with larger cells. Conf Apic Trop Climate 2: 238-253. New Delhi

Jander R, Jander K (19709 Über die Phylogenie der Geotaxis innerhalb der Bienen (*Apoidea*). Z vgl Physiol 66: 355-368

Jay SG (1963) The development of honeybees in their cells. J Apic Res 2: 117-134

Joshi MA, Diwan VV, Suryanarayana MC (1980) Bees and honey in ancient India. Int Conf Apicult Trop Climate, New Delhi 2: 143-149

Kallapur SK (1950) An experiment in the collection of wild honey. Indian Bee J 12: 122-124

Kalman Ch (1973) Our formerly very aggressive bee. Proc Int Beekeep Congr 24: 306-307

Kapil RP (1956) Variation in biometrical characters of the Indian honeybee (*Apis indica* F). Indian J Entomol 18: 440-457

Kapil RP (1957) The length of life and the brood-rearing cycle of the Indian bee. Bee World 38: 258-263

Kapil RP (1960) Osservazioni sulla temperatura del glomere dell'ape indiana. Apic Ital 27: 79-83

Kapil RP (1962) Anatomy and histology of the female reproductive system of *Apis indica*. Ins sociaux 9: 145-153

Kapil RP, Jain KL (1980) Biology and utilization of insect pollinators for crop production. Dept Zool Haryana Agr Univ Hissar, India

Kaschef AH (1959) The single strain of the Egyptian honeybee. Ins Sociaux 6: 243-257

Kauhausen D (1984) Genetische Analyse der telythoken Parthenogenese bei der Kap-Honigbiene (*Apis mellifera capensis*). Thesis, FB Biology, University, Frankfurt

Kellog CR (1941) Some characteristics of the Oriental honeybee, *Apis indica* F in China. J econ Entomol 34: 717-779

Kellog CR (1967) Entomological excerpts from southeastern China. Claremont, CA

Kellog CR (1972) Chinese-Japanese bees. In Root Th, The ABC and XYZ of Bee Culture. Root Corp, Medina OH

Kelner-Pillault S (1969) Abeilles fossiles ancêtres des Apides sociaux. Proc Int IUSSI Congr IV: 85-93

Kerr WE, Bueno D (1970) Natural crossing between *Apis mellifera adansonii* and *Apis mellifera ligustica*. Evol 24: 145-148

Kerr WE, da Cunha RA (1976) Taxonomic position of two fossil social bees (*Apidae*) Rev Biol Trop 24: 35-43

Kerr WE, Laidlaw H (1956) General genetics of bees. Advan Genet 8: 109-153

Kerr WE, Maule V (1964) Geographic distribution of stingless bees and its implications (*Hymenoptera: Apidae*). J Entomol Soc NY 72: 2-17

Kerr WE, de Portugal Araujo V (1958) Racas de abelhas de Africa. Garcia de Ortä 6: 53-59

Kerr WE, Sakagami SF, Zucchi R, Portugal Araujo V, Camargo JM (1967) Onservacoes sobre a arquitetura dos ninhos e comportamento de algumas espécies de abelhas sem ferrao das vizinhancas de Manaus, Amazonas (*Hymenoptera, Apoidea*). At Simp Biota Amazonica 5: 255-309

Kevan PG, Morse RA, Akranatakoul P (1984) Apiculture in tropical and subtropical Asia with special reference to European honeybee development programmes. Proc Expert Consult Bangkok 10-33. FAO Rome

Khan AH, Afzal M (1950) Vicinism in cotton. Indian Cotton Grow Rev 4: 227-239

Kiesenwetter EAH (1860) Über die Bienen des Hymettos. Berlin Entomol Nachr 1860: 315-317

Kigatiira KI (1979) Behaviour of the East African honeybee. Proc Int Beekeep Congr 27: 295-299

Kigatiira KI (1984) Bees and beekeeping in Kenya. Bee World 65: 74-80

Kimsey LS (1984) A re-evaluation of the phylogenetic relationships in the Apidae (Hymenoptera). Syst Entomol 9: 435-441

Koeniger G (1986) Reproduction and mating behavior. In Rinderer TE (ed) Bee Breeding and Genetics, 255-288. Academic Press, Orlando FA

Koeniger N (1975) Observations on alarm behaviour and colony defense of *Apis dorsata*. IUSSI Symp Pheromones Def Secretions Soc Ins 153-154

Koeniger N (1976a) The Asiatic honeybee *Apis cerana*. Proc Apic Trop Climate 47-49. IBRA Gerrards Cross

Koeniger N (1976b) Neue Aspekte der Phylogenie innerhalb der Gattung *Apis*. Apidologie 7: 257-266

Koeniger N (1977) Keeping the three *Apis* species from Asia, *Apis cerana, Apis dorsata, Apis florea* in bee flight rooms. Ins Sociaux 24: 286-289

Koeniger N (1982) Interactions among the four species of the genus *Apis*. In: Breed Michener Evans (ed) The biology of social insects. Westview Press, Boulder CO

Koeniger N, Delfinado-Baker M (1983) Observation on mites of the Asian honeybee species *(Apis cerana, Apis dorsata, Apis florea)*. Apidologie 14, 197-204

Koeniger N, Fuchs S (1973) Sound production as colony defense in *Apis cerana* Fabr. Proc Int IUSSI Congr 7: 199-204

Koeniger N, Fuchs S (1975) Zur Kolonieverteidigung östlicher Honigbienen. Z Tierpsych 37: 99-106

Koeniger N, Koeniger G (1980) Observations and experiments on migration and dance communication of *Apis dorsata* in Sri Lanka. J Apic Res 19: 21-34

Koeniger N, Vorwohl G (1979) Competition for food among four sympatric species of Apini in Sri Lanka (*Apis dorsata, Apis cerana, Apis florea* and *Trigona irridipennis*) J Apic Res 18: 95-109

Koeniger N, Wijayagunasekera HNP (1976) Time of drone flight in the three Asiatic honeybee species (*Apis cerana, Apis florea, Apis dorsata*). J Apic Res 15: 67-71

Koeniger N, Weiss J, Ritter W (1975) Catch, transport and flight cage maintenance of the Giant honeybee *Apis dorsata*. Proc Int Beekeep Congr 25: 313-317

Koeniger N, Weiss J, Maschwitz U (1979) Alarm pheromones of the sting in the genus *Apis*. J Insect Physiol 25: 467-476

Koeniger N, Koeniger G, Wijayagunasekera HNP (1980) Beobachtungen über die Anpassung von *Varroa jacobsoni* an ihren natürlichen Wirt *Apis cerana* in Sri Lanka. Apidologie 12: 37–40

Koeniger N, Koeniger G, Punchihewa RKW, Fabritius Mo, Fabritius Mi (1982) Observations and experiments on dance communication in *Apis florea* F. J Apic Res 21: 45–52

Koivulehto K (1974) Beekeeping in Finland. Symp Beekeep cold zones, Sci Bull Apimondia, 70–72

Koltermann R (1973 a) Retroaktive Hemmung nach successiver Informationseingabe bei *Apis mellifera* und *Apis cerana* (*Apidae*). J Comp Physiol 84: 299–310

Koltermann R (1973 b) Rassen- bzw. artspezifische Duftbewertung bei der Honigbiene und ökologische Adaptation. J Comp Physiol 85: 327–360

Kotzin Th (1931) Beekeeping in Transbaikalia. Bee World 12: 78

Kreil G (1973) Structure of melittin isolated from two species of honeybees. FEBS Letters 33: 241–244

Kreil G (1975) The structure of *Dorsata* melittin: phylogenetic relationships between honeybees as deducted from sequence data. FEBS Letters 54: 100–102

Kshirsagar KK (1980) Morphometric studies on the Indian hive bee *Apis cerana indica* F. I. Morphometric characters useful in identification of intraspecific taxa. Proc Int Conf Apic Trop Climate 2: 254–261. New Delhi

Kshirsagar KK (1982) Current incidence of honeybee diseases and parasites in India. Bee World 63: 162–164

Kshirsagar KK, Muvel KS, Mittal MC, Phadke RP (1980) Some observations on behaviour of *Apis florea* F. Proc Int Conf Apic Trop Climate 2: 356–366. New Delhi India

Kuss SE (1973) Die pleistozänen Säugetierfaunen der ostmediterranen Inseln, ihr Alter und ihre Herkunft. Ber naturforsch Ges Freiburg Br 63: 49–71

Kuss SE (1975) Die pleistozänen Hirsche der ostmediterranen Inseln Kreta, Kasos, Karpathos und Rhodos (Griechenland). Ber naturf Ges Freiburg Br 65: 25–79

Laidlaw HH (1987) The history of instrumental insemination of honeybee queens. Bee World 68 (in press)

Landa Friar D de (1566) Yucatan. Before and after the conquest. Reprint Dover Publ New York 1978

Lanerolle GA (1984). See FAO 1984

Latif A, Qayyum A, Manzoor-ul-Haq (1956) Researches on the composition of Pakistan honey. Pak J Sci Res 8: 163–166

Latif A, Qayyum A, Manzoor-ul-Haq (1960) Multiple and two-queen systems in *Apis indica* F. Bee World 41: 201–209

Latreille PA (1804) Notice des espèces d'abeilles vivant en grande societé, et formant de cellules hexagonales, ou des abeilles proprement dites. Ann Mus Nat Hist Natur Paris 5: 161–178

Lattin G de (1967) Grundriß der Zoogeographie. Fischer, Stuttgart

Lauer J, Lindauer M (1971) Genetisch fixierte Lerndispositionen bei der Honigbiene. Abh Akad Wiss Mainz Inf Org 1: 1–87

Lauer J, Lindauer M (1973) Beteiligung von Lernprozessen bei der Orientierung der Honigbiene. Fortschr Zool 21: 349–370

Lavrekhin FF (1958) On the first attempts to introduce the wild Ussurian bees (*Apis indica* F.) to the European part of the USSR. Proc Int Beekeep Congr 17: 237–238

Lepeletier A (1836) Histoire naturelle des insectes. Hymenoptères 1: 400–407. Roret, Paris

Lepissier J (1968) L'apiculture en Republique Centrafricaine. Rep Min Dev Bangui

Leporati M, Valli M, Cavicchi S (1984) Etude biométrique de la variabilité géographique des populations d'*Apis mellifera* en Italie septentrionale. Apidologie 15: 285–302

Leuenberger F, Morgenthaler O (1954) Die Biene. 3. Auf Sauerländer, Aarau

Lindauer M (1956) Über die Verständigung bei indischen Bienen. Z vgl Physiol 38: 521–557

Lindauer M (1957) Communication among the honeybees and stingless bees of India. Bee World 38: 3–14, 34–39

Lindauer M, Kerr WE (1960) Communication between the workers of stingless bees. Bee World 41: 29–41, 65–71

Linnaeus C (1758) Systema Naturae. 10th edn. Holmiae Laur Salvii

Linnaeus C (1761) Fauna Suecica. Stockholmiae Laur Salvii

Li Shawen, Meng Yupin, Chang JT, Li Juhnai (1987) Comparative study of esterase isozymes of six species of honeybees. Proc Int Beekeep Congr 31 (in press)

Longo S (1980) Consistenza attuale e prospetti di sviluppo dell'apicoltura nei Monti Iblei. Atti 3 Convegno Siciliano di Ecologia. Iblei: La natura e l'uomo 1-13. Noto

Longo S (1982) Indagine sulla consistenza dell'apicoltura nella Sicilia orientale. Apic Moderno 73: 11-21

Longo S (1984) L'apicoltura in Scilia orientale. Stato attuale e prospettive di sviluppo. Apitalia 1984: 1-7

Longo S, Mauro S, Coco A (1980). Indagine conoscitiva su alcuni aspetti dell' apicoltura siciliana. Tec Agric 32: 5-17

Louis J (1963) Etude de la translation discoidale de l'aile de l'abeille. Ann Abeille 6: 303-320

Louis J (1973) La nomenclature d'aile des hymenoptères. Essai de normalisation. Beitr Entomol 23: 276-289

Louveaux J (1969) Ecotype in honeybees. Proc Int Beekeep Congr 22: 499-501

Louveaux J, Albisetti M, Delangue M, Theurkauff M (1966) Les modalités de l'adaptation des abeilles (*Apis mellifica*) au milieu naturel. Ann Abeille 9: 323-350

Lunder R (1953) Foraging characteristics of some bee races in Norwegian conditions. Bitt Nord 3: 71-83 (Norwegian)

Maa TC (1953) An inquiry into the systematics of the Tribus Apidini or honeybees (*Hymenoptera*). Treubia 21: 525-640

McGregor SA (1976) Insect pollination of cultivated crop plants. Agric Hand 496, USDA Washington DC

Mackensen O (1943) The occurrence of parthenogenetic females in some strains of honeybees. J Econ Entomol 36: 465-467

Ma De-Feng, Huang Wen-Cheng (1981) Apiculture in the New China. Bee World 62: 163-166

Mahindra DB, Muvel KS, Kshirsagar KK, Thakar CV, Phadke RP, Deodikar GB (1977) Nesting behaviour of *Apis dorsata*. Proc Int Beekeep Congr 26: 299-300

Makhdzir M (1984) See FAO 1984

Makhdzir M, Osman S (1980) Beekeeping in coconut smallholdings in Pontian, Johor, West Malaysia. Proc Int Conf Apic Trop Climate 2: 179-186. New Delhi

Mandel G (1980) Das Reich der Königin von Saba. Scherz, Bern

Markosjan AA, Akopjan NM, Abgarov GB (1976) Characteristics of the yellow Armenian bee. Proc Apimondia Symp Genetics Selection Reproduction: 173-175

Marletto F, Manino A, Pedrini P (1984a) Intergradazione fra sottospezie di *Apis mellifera* in Liguria. Apic Moderno 75: 159-163

Marletto F, Manino A, Pedrini P (1984b) Indagini biometriche su populazione di *Apis mellifera* L delle Alpi occidentali. Apic Moderno 75: 213-223

Matsuura M, Sakagami SF (1973) A bionomic sketch of the Giant hornet, *Vespa mandarinia*, a serious pest for Japanese apiculture. J Fac Sci Hokkaido Univ Ser VI Zool 19: 125-162

Mattu VK, Verma LR (1980) Comparative morphometric studies on the introduced European bee *Apis mellifera* L and Indian honeybee *Apis cerana indica* F in Himachal Pradesh. Int Conf Apic Trop Climate 2: 262-277. New Delhi, India

Mattu VK, Verma LR (1983) Comparative morphometric studies on the Indian honeybee of the north-west Himalayas. 1.Tongue and antenna. J Apic Res 22: 79-85

Mattu VK, Verma LR (1984a) Morphometric studies on the Indian honeybee *Apis cerana indica*. Effect of seasonal variations. Apidologie 15: 63-73

Mattu VK, Verma LR (1984b) Comparative morphometric studies on the Indian honeybee of north-west Himalayas. 2.Wings. J Apic Res 23: 1-10

Maun GS, Gurdip S (1983) Activity and abundance of pollinators of plums at Ludhiana (Punjab) Am Bee J 123: 595

Mayr E (1963) Animal species and evolution. Harvard Univ Press Cambridge Mass

Meer Moor JC van der (1932) Aanval op bijennest door *Pernis apivorus*. Trop Natuur 21: 67-68

Menzel R, Freudel H, Rühl U (1973) Rassenspezifische Unterschiede im Lernverhalten der Honigbiene (*Apis mellifera* L). Apidologie 4: 1-24

Mestriner MA (1969) Biochemical polymorphisms in bees *(Apis mellifera ligustica)*. Nature (Lon) 223: 188-189

Michailov AS (1924) On the tongue length of the bees from Tscherepovetz district in connection with the problems of local differences in tongue length. Pchelowodnoje Djels, February (Russian)

Michailov AS (1926) Über lineare Korrelation zwischen Rüssellänge der Honigbiene und der geographischen Breite im ebenen europäischen Rußland. Arch Bienenkd 7: 28–33

Michailov AS (1927a) Der Einfluß einiger Lebenslagefaktoren auf die Variabilität der Honigbiene *(Apis mellifera)*. Arch Bienenkd 8: 289–303

Michailov AS (1927b) Über die Saison-Variabilität der Honigbiene. Arch Bienenkd 8: 304–312

Michailov AS (1927c) Über den Zusammenhang zwischen dem Umfang der Bienenzelle und dem Umfang des Bienenkörpers und seiner Teile. Arch Bienenkd 8: 313–321

Michailov AS (1928) Effect of the colony upon the bees reared in it. Opytnaje Paseca 7: 299–302 (Russian)

Michailov AS (1930) A summary of what we know of long-tongued bees. Am Bee J 70: 532–533

Michener CD (1944) Comparative external morphology, phylogeny and classification of the bees (Hymenoptera). Bull Am Mus Nat Hist 82: 151–326

Michener CD (1974) The social behavior of the bees. Harvard Univ Press

Michener CD (1975) The Brazilian bee problem. Annu Rev Entomol 20: 399–416

Millen TW (1942a) Bee breeding, laying workers and their progeny. Indian Bee J 4: 94–95

Millen TW (1942b) *Apis dorsata* queens. Indian Bee J 4: 6–9

Mishra RC, Dogra GS (1983) Post-embryonic development of *Apis cerana indica* F worker bee. Int Conf Apic Trop Climate 2: 278–288

Miyamoto S (1958) Biological studies on Japanese bees. X Differences in flower relationship between a Japanese and a European honeybee. Sci Rep Hyogo Univ Agric Ser agric Biol: 99–101

Montagano J (1911) Relation sur l'*Apis sicula*. Proc Int Beekeep Congr 5: 26–29

Moritz RFA, Hänel H (1984) Restricted development of the parasitic mite *Varroa jacobsoni* Oudemans in the Cape honeybee *Apis mellifera capensis* Esch. Z angew Entomol 97: 91–95

Moritz RFA, Kauhausen D (1984) Hybridization between *Apis mellifera capensis* and adjacent races of *Apis mellifera*. Apidologie 15: 21–222

Moritz RFA, Klepsch A (1985) Estimating heritabilities of worker characters: a new approach using laying workers of the Cape honeybee *(Apis mellifera capensis)*. Apidologie 16: 47–56

Morse RA (1969) The biology of *Apis dorsata* in the Philippines. Proc Int IUSSI Congr 6: 165–167

Morse RA (1970a) Annotated bibliography on *Apis dorsata*. IBRA Gerrards Cross

Morse RA (1970b) Annotated bibliography on *Apis florea*. IBRA Gerrards Cross

Morse RA, Benton AW (1967) Venom collection from species of honeybees in South-East Asia. Bee World 48: 19–29

Morse RA, Laigo FM (1969) *Apis dorsata* in the Philippines. Monogr Philipp Ass Entomol 1: 1–96

Morse RA, Shearer DA, Boch R, Benton AW (1969) Observations on alarm substances in the genus *Apis*. J Apic Res 6: 113–118

Mukwaira B (1976) Traditional and transitional beekeeping in Rhodesia. Proc Apimondia Symp Afr Bees 103–106

Muttoo RN (1951) The correct scientific nomenclature for our Indian hive bees. Indian Bee J 13: 150–153

Muttoo RN (1956) Facts about beekeeping in India. Bee World 37: 125–133; 154–157

Narayanan ES, Sharma PL, Phadke KG (1961) Studies on the biometry of the Indian bees III Tongue length and number of hooks on the hind wings of *Apis indica* F collected from Madras. State Indian Bee J 23: 3–9

N'Diaye M (1974) L'apiculture au Senegal. Thèse, Fac Mèd Pharm, Ecole veterinaire Dakar

Nightingale JM (1976) Traditional beekeeping among Kenyan tribes and methods proposed for improvement and modernization. Proc Conf Apic trop climate 15–22 IBRA Gerralds Cross

Nogge G (1974) Die geographische Verbreitungsgrenze zwischen westlicher und östlicher Honigbiene. Allg Dtsch Imkerztg 8: 163–165

Nordenskioeld E (1934) Beekeeping among the American Indians. Bee World 15: 26–28

Nunamaker RA, Wilson WT, Haley BEC (1984) Electrophoretic detection of Africanized honeybees (*Apis mellifera scutellata*) in Guatemala and Mexico based on malate dehydrogenase allozyme patterns. J Kans Ent Soc 57: 622–631

Nunes JRR, Tordo GC (1966) Prospeccoes e ensaios experimentais apicolas em Angola. Lisbon Junta Invest Ultramar Estud Ens Doc 70: 137-159

Okada I (1970) About the Japanese honeybee. Studies on honeybees 32: 259-282 Tamagawa University, Tokyo (Japanese)

Okada I (1985) Biological characteristics of the Japanese honeybee, *Apis cerana japonica*. Proc Int Apic Congr 30: 119-122

Okada I, Sakai T (1960) A comparative study on natural comb of Japanese and European honeybee, with special difference in cell number. Bull Fac Agr Tamagawa Univ 1: 1-11

Okada I, Sakai T, Hasegava M (1956) Notes on some morphological characters of Japanese honeybees. Kontyu 24: 145-154

Okada I, Sakai T, Obata H (1958) On the habits of the Japanese honeybees. Proc Int Beekeep Congr 17: 243-244

Oldroyd B, Moran C (1983) Heritability of worker characters in the honeybee *(Apis mellifera)*. Aust J of Biol Sci 36: 323-332

Onions GW (1912) South African "fertile-worker bees". Apric J Un S Afr 3: 720-728

Oschmann H (1961) Eine bienenkundliche Reise in die Volksrepublik China. Arch Geflügelz Kleintierkd 10: 235-259

Pandey RS (1977) Behavior of the Indian Honey Bee in double brood chamber hives. Am Bee J 117: 627

Phadke RP (1961) Some physico-chemical constants of Indian beeswax. Bee World 42: 149-153

Peng Y-S, Fang Y, Xu Sh, Ge L (1987) Resistance mechanism of the Asiatic honeybee *Apis cerana* Fabr to an ectoparasitic mite *Varroa jacobsoni* Oud. J Invert Path 49: 54-60

Pollmann A (1889) Wert der verschiedenen Bienenrassen und deren Varietäten. 2nd edn Voigt, Berlin Leipzig (1st edn with description of *A. m. carnica* 1879)

Portugal Araujo de V (1956) Notas bionomicas sobre *Apis mellifera adansonii* Latr. Dusenia 7: 91-102

Prota R (1976) Osservazioni sulla variabilità somatometrica delle populazioni sarde di *Apis mellifera ligustica* spinola. L'Apic moderno Torino 67: 77-81

Punchihewa RWK, Koeniger N, Kevan PG, Gadawski RM (1985) Observations on the dance communication and natural foraging ranges of *Apis cerana, Apis dorsata* and *Apis florea* in Sri Lanka. J Apic Res 24: 168-175

Qayyum HA, Nabi A (1968) Biology of *Apis dorsata* F. Pak J Sci 19: 109-113

Radoszkowski OI (1877) Hymenoptères de Korée. Hor Soc entomol Ross Petersburg 21: 428-436

Rafig A (1984) Country status report on beekeeping in Pakistan. Proc Expert Consult 203-210 FAO Rome

Rahman KA (1945) Progress of beekeeping in the Punjab. Bee World 26: 42-44; 50-52

Rahman KA (1947) Size of comb cells of the Indian Bee. Bee World 28: 68-69

Rahman KA, Singh S (1950) Variation in the tongue length of the honeybees. Indian J Entomol 10: 63-73

Rashad SE, El-Sarrag MS (1980) Some characters of the Sudanese honeybee *Apis mellifera L.* Int Conf Apic trop climate 2: 301-309 New Delhi

Rashad SE, El Sarrag MS (1984) Beekeeping in Sudan. Two morphometrical studies on the Sudanese Honeybees. Unpubl manuscript

Reddy Ch (1980a) Observations on the annual cycle of foraging and brood rearing by *Apis cerana indica* colonies. J Apic Res 19: 17-20

Reddy Ch (1980b) Studies on the nesting behavior of *Apis dorsata* F. Int Conf Apic trop climate 2: 391-397 New Delhi Inst Agr Res

Renner S (1982) Die Bienen Amazoniens - potentielle Honiglieferanten? Allg. Forstzg 37: 938-939

Rensch B (1929) Das Prinzip geographischer Rassenkreise und das Problem der Artbildung. Borntraeger, Berlin

Rensch B (1936) Studie über klimatische Parallelität der Merkmalsausprägung bei Vögeln und Säugern. Arch Naturgesch N F 5: 317-363

Rensch B (1939) Klimatische Auslese von Größenvarianten. Arch Naturgesch N F 8: 98-128

Rensch B (1950) Klima und Artbildung. Geolog Rundschau 38: 137-152

Riecke-Lauer J, Lindauer M (1985) Lernprozesse im Orientierungsablauf der Honigbiene. Ein rassenspezifischer Vergleich von *Apis mellifera carnica* und *Apis mellifera ligustica*. Fischer, Stuttgart New York

Rinderer TE (1982) Behavioral genetic analysis of colony defence by honeybees. Proc Soc Ins Trop 1: 249–254

Rinderer TE, Collins AM, Bolten AB, Harbo JR (1981) Size of nest varieties selected by swarms of Africanized honeybees in Venezuela. J Apic Res 20: 160–164

Rinderer TE, Hellmich II RL, Danka RG, Collins AM (1985) Male reproductive parasitism: a factor in the Africanization of European honey-bee populations. Science 228: 1119–1121

Rinderer TE, Sylvester HA, Collins AM, Pesante D (1986a) Identification of Africanized and European honey bees: effects of nurse-bee genotype and comb size. Bull Entomol Soc Amer 32: 150–152

Rinderer TE, Sylvester HA, Brown MA Villa JD, Pesante D, Collins MA (1986b) Field and simplified techniques for identifying Africanized and European bees. Apidologie 17: 33–48

Rinderer TE, Sylvester HA, Buco SM, Lancaster VA, Herbert EW, Collins AM, Helmich II RL (1987a) Improved simple techniques for identifying Africanized and European honeybees. Apidologie 18 (in press)

Rinderer TE, Collins A, Sylvester A (1987b) Differential drone production by Africanized and European honeybee colonies. Apidologie 18 (in press)

Roberts WC, Mackensen O (1951) Breeding improved honeybees. II Heredity and variation. Am Bee J 91: 328–330

Robinson WS (1981) Beekeeping in Jordan. Bee World 62: 91–97

Roegl F, Steininger F (1983) Vom Zerfall der Tethys zu Mediterran und Paratethys. Ann Naturhist Mus Wien 85A: 135–163

Roegl F, Steininger F (1984) Neogene Paratethys, Mediterranean and Indo pacific seaways. In: (ed. P. Brenchley) Fossils and Climate, pp 171–200. J. Wiley Sons Ltd

Roepke W (1930) Beobachtungen an indischen Honigbienen, insbesondere an *Apis dorsata*. Meded Landbouwhooge-school Wageningen 34: 1–28

Roswall G (1974) Overwintering and honey yields in the polar circle region of Sweden. Symp Apic in cold zones 93–95 Apimondia, Bucharest

Rothe U, Nachtigall W (1980) Zur Haltung von *Apis mellifica* in einem Flugraum. Apidologie 11: 17–24

Rothenbuhler WC (1964) Behaviour genetics of nest cleaning in honeybees. I Responses of four inbred lines to disease killed brood. Anim behav 12: 578–583

Rothenbuhler WC, Kulincevic JM, Kerr WE (1968) Bee genetics. Ann Rev Genet 2: 413–437

Rotter E (1920) The Egyptian bee (*Apis mellifica* var *fasciata*). Bee World 2: 76–81

Rotter E (1921) Die ägyptische Biene. Arch Bienenkd 3: 1–8

Roubik DW (1983) Nest and colony characteristics of stingless bees from Panama (*Hymenoptera*: *Apidae*). J Kansas Entomol Soc 56: 327–355

Roubik DW, Ahja M (1983) Flight ranges of *Melipona* and *Trigona* in tropical forests. J Kansas Entomol Soc 56: 217–222

Roubik DW, Peralta FJA (1983) Thermodynamics in nests of two *Melipona* species in Brazil. Acta Amazonica 13: 453–466

Roubik DW, Sakagami ShF, Kudo I (1985) A note on distribution and nesting of the Himalayan honeybee *Apis laboriosa* Smith (*Hym*: *Apidae*) J Kansas Entomol Soc 58: 746–749

Ruttner F (1950) Untersuchungen über die Bienenrassen Österreichs. Bienenvater (Wien) 75: 168–176

Ruttner F (1952a) Alter und Herkunft der Bienenrassen Europas. Oesterr Imker 2: 8–10

Ruttner F (1952b) Zur Dyssymmetrie des Flügelgeäders der Honigbiene. Z Bienenforsch 1: 219–224

Ruttner F (1952c) Die Außenmerkmale des *Carnica*-Stammes Troiseck. Oesterr Imker 2: 67–69

Ruttner F (1953) Über die Vererbung einiger Rassenmerkmale bei der Honigbiene (*Apis mellifera*). Oesterr Zool Z 4: 183–190

Ruttner F (1954a) Rassentypische Unterschiede in der Orientierungsfähigkeit. Oesterr Imker 4: 139–142

Ruttner F (1954b) Die Heimat der Carnica. Dtsch Bienenwirtsch 5: 14–16; 28–31; 54–57; 63–67; 109–110

Ruttner F (1957a) Aktuelle Probleme auf dem Gebiet der Fortpflanzung und Züchtungsforschung der Biene. Dtsch Bienenwirtsch 8: 41–44

Ruttner F (1957b) Die Sexualfunktionen der Honigbienen im Dienste ihrer sozialen Gemeinschaft. Z vgl Physiol 39: 577–600

Ruttner F (1965) Versuch einer Charakterisierung der *Carnica*-Biene nach ihrem Flügelgeäder. Ustav Vedeckotech Inf MZLVH: 165–172. Praha

Ruttner F (1969) Biometrische Charakterisierung der österreichischen *Carnica*-Biene. Z Bienenforsch 9: 469–491

Ruttner F (1973a) Die Bienenrassen des Mediterranen Beckens. Apidologie 4: 171–172

Ruttner F (1973b) Drohnen von *Apis cerana* Fabr auf einem Drohnensammelplatz. Apidologie 4: 41–44

Ruttner F (1975a) Die Kretische Biene, *Apis mellifera adami*. Allg dtsch Imkerztg 9: 271–272

Ruttner F (1975b) Ein metatarsaler Haftapparat bei den Drohnen der Gattung *Apis* (Hymenoptera, Apidae). Entomol German 2: 22–29

Ruttner F (1976a) Beekeeping in Crete. Apiacta 11: 187–191

Ruttner F (1976b) Isolated populations of honeybees in Australia. J Apic Res 15: 97–104

Ruttner F (1976c) African races of honeybees. Proc Int Beekeep Congr 25: 325–344

Ruttner F (1977a) The Cape bee: a biological curiosity. Proc Apimondia Symp Afr Bees: 127–131

Ruttner F (1977b) The problem of the Cape bee (*Apis mellifera capensis* Escholtz): parthenogenesis, size of population, evolution. Apidologie 8: 281–294

Ruttner F (1980a) *Apis mellifera adami* (n ssp) Die Kretische Biene. Apidologie 11: 385–400

Ruttner F (1980b) L'abeille du Tell et son utilization dans l'apiculture moderne. Cah Rech 11: 32–38. Curer Constantine

Ruttner F (1981) Taxonomy of honeybees of tropical Africa. Proc Int Beekeep Congr 28: 271–277

Ruttner F (ed) (1983a) Queen rearing. Apimondia, Bucharest

Ruttner F (1983b) Zuchttechnik und Zuchtauslese. 5.Aufl, Ehrenwirth, München

Ruttner F (1985a) Graded geographic variability in honeybees and environment. Pszczeln Zeszyty Nauk 29: 81–92. Pulawy, Poland

Ruttner F (1985b) Reproductive behavior in honeybees. Fortschr Zool 31. Hölldober B, Lindauer M (eds) Exper Behav Ecol Sociobiol, 225–236

Ruttner F (1986a) Geographic variation and taxonomy of *Apis cerana*. Proc Int Beekeep Congr 30: 130–133

Ruttner F (1986b) Geographical variability and classification. In Th Rinderer (ed.) Bee Genetics and Breeding: 23–56 Academic Press Orlando FA

Ruttner F (1987) Breeding techniques and selection for breeding of the honeybee. Northern Bee Books, Mytholmroyd, UK

Ruttner F, Hesse B (1979) Rassenspezifische Unterschiede in Ovarentwicklung und Eiablage von weisellosen Arbeiterinnen der Honigbiene *Apis mellifera* L Apidologie 12: 159–183

Ruttner F, Kaißling KE (1968) Über die interspezifische Wirkung des Sexuallockstoffes von *Apis mellifera* und *Apis cerana*. Z vgl Physiol 59: 362–270

Ruttner F, Kauhausen D (1985) Honeybees of tropical Africa: ecological diversification and isolation. Proc Int Conf Apic Trop Climate 3: 45–51

Ruttner F, Maul V (1983) Experimental analysis of the reproductive interspecific isolation of *Apis mellifera* L. and *Apis cerana* Fabr. Apidologie 14: 309–327

Ruttner F, Woyke J, Koeniger N (1972) Reproduction in *Apis cerana*. 1 Mating behaviour. J Apic Res 11: 141–146

Ruttner F, Woyke J, Koeniger N (1973) Reproduction in *Apis cerana*. 2 Reproductive organs and natural insemination. J Apic Res 12: 21–34

Ruttner F, Koeniger N, Veith HJ (1976) Queen substance bei eierlegenden Arbeiterinnen der Honigbiene *Apis mellifera* L. Naturwissenschaften 63: 434

Ruttner F, Tassencourt L, Louveaux J (1978) Biometrical-statistical analysis of the geographic variability of *Apis mellifera* l. 1 Material and methods Apidologie 9: 363–381

Ruttner F, Pourasghar D, Kauhausen D (1985a) Die Honigbienen des Iran 1 *Apis florea* Fabr. Apidologie 16: 119–138

Ruttner F, Pourasghar D, Kauhausen D (1985b) Die Honigbienen des Iran 2 *Apis mellifera meda* Skor, die Persische Biene. Apidologie 16: 241–264

Ruttner F, Wilson R, Snelling G, Vorwohl G, Kauhausen D (1986) The evolution of the honeybee wing venation. Apidologie 17: 349

Sakagami ShF (1958) Zum gegenwärtigen Zuchtzustand japanischer Honigbienen in Kiushu, Süd-Japan. Z Bienenforsch 4: 87–96

Sakagami ShF (1959) Some interspecific relations between Japanese and European honeybees. J Anim Ecol 28. 51–58

Sakagami ShF (1960a) Preliminary report on the specific difference of behavior and other ecological characters between European and Japanese honeybees. Acta Hymenopterologica 1: 171–198

Sakagami ShF (1960b) Two opposing adaptations in the post-stinging response of the honeybees. Studies on the Japanese honeybees VIII. Evol 14: 29–40

Sakagami ShF (1971) Ethosoziologischer Vergleich zwischen Honigbienen und Stachellosen Bienen. Z Tierphysiol 28: 337–350

Sakagami ShF, Akahira Y (1958) Comparison of ovarian size and number of ovarioles between workers of Japanese and European honeybees (Studies on the Japanese honeybee *Apis indica cerana* F I) Kontyu 26: 103–107

Sakagami ShF, Konta Sh (1958) An attempt to rear the Japanese bee in a framed hive. J Fac Sci Hokkaido Univ Ser VI Zool 14: 1–8

Sakagami ShF, Yoshikawa (1973) Additional observation on the nest of the dwarf honeybee *Apis florea.* Kont 41: 217–219

Sakagami ShF, Matsumura T, Ito K (1980) *Apis laboriosa* in Himalaya. The little known worlds largest honeybee *(Hymenoptera, Apidae).* Insecta Matsumurana NS 19: 47–77

Sakai T (1956) Morphological studies on the difference among some strains of the honeybee. Jpn Bee J 9: 1–11

Sakai T, Matsuka M (1982) Beekeeping and honey production in Japan. Bee World 63: 63–71

Sandhu As, Singh S (1966) The biology and brood rearing activities of the little honeybee (*Apis florea* Fabr). Indian Bee J 22: 27: 34

Sannasi A, Rajulu G, Sundara G (1971) 9-oxo-trans-2-decenoic acid in the Indian honeybee. Life Sci 10: 195–201

Santiago E, Albornoz J, Dominguez A, Inquierdo JI (1986) Etude biométrique des populations d'abeilles (*Apis mellifera*) du nord-ouest de l'Espagne. Apidologie 16: 71–92

Saraf FK, Wali JL (1972) Preliminary studies on the egg laying capacity of the queen bee of *Apis indica* F (hill strain) in Kashmir. Indian Bee J 34: 27–31

Saudi Arabia (1980) Statistical year book 16. Min Fin Nat Econ Riyadh

Schaffer WM, Zeh DW, Schaffer MW (1982) Competition among social species. Proc Int IUSSI Congr 9: 14–18

Schmidt JO, Schmidt PJ, Starr CK (1985) Investigating the Giant honeybee, *Apis dorsata,* in Sabah. Am Bee J 125: 749–751

Schneider D (1971) A "do it yourself" bee operation. Am Bee J 111: 462–463

Schneider P, Djalal AS (1970) Vorkommen und Haltung der Östlichen Honigbiene (*Apis cerana* Fabr) in Afghanistan. Apidologie 1: 329–341

Schneider P, Kloft W (1971) Beobachtungen zum Gruppenverteidigungsverhalten der Östlichen Honigbiene *Apis cerana* Fabr. Z Tierpsych 29: 337–342

Schricker B (1981) Kurzbericht über Untersuchungen zur Minderung der Aggressivität der Zentralafrikanischen Honigbiene (*Apis mellifera adansonii*) durch synthetisches Nasanoff-Pheromon. Apidologie 12: 94–95

Schwarzbach M (1974) Das Klima der Vorzeit. Enke, Stuttgart

Seeley TD (1978) Life history strategy of the honey bee, *Apis mellifera.* Oecologia (Berlin) 32: 109–118

Seeley TD (1982) How honeybees find a home. Scientific American 247: 158–168

Seeley TD (1983) The ecology of temperate and tropical honeybee societies. Am Sci 71: 264–272

Seeley TD (1985) Honeybee ecology. Princeton Univ Press, New Yersey

Seeley TD, Heinrich B (1981) Regulation of temperature in the nests of social insects. In: Heinrich B (ed) Insect Thermoregulation 159–234, Wiley, NY

Seeley TD, Morse RA (1976) The nest of the honeybee. Ins Sociaux 23: 495–512

Seeley TD, Morse RA (1977) Dispersal behavior of honeybee swarms. Psyche 84: 199–209

Seeley TD, Visscher PK (1985) Survival of honeybees in cold climes: the critical timing of colony growth and reproduction. Ecol Entomol 10: 81-88

Seeley TD, Seeley RH, Akratanakul P (1982) Colony defense strategies of the honeybee in Thailand. Ecol Monogr 52: 43-63

Shah AM (1980) Beekeeping in Kashmir. Proc Int Conf Apic Trop Climate 2: 197-204, New Delhi

Shah FA, Shah TA (1980a) Flight range of *Apis cerana* from Kashmir. Indian Bee J 42: 48

Shah FA, Shah TA (1980b) Early life, mating and egg laying of *Apis cerana* queens in Kashmir. Bee World 61: 137-140

Sharma OP, Mishra RC, Dogra GS (1980) Management of *Apis mellifera* in Himachal Pradesh. Proc Int Conf Apic Trop Climate 2: 205-210

Sharma PL (1960a) Observations on the swarming and mating habits of Indian honeybee. Bee World 41: 121-123

Sharma PL (1960b) Experiments with *Apis mellifera* in India. Bee World 41: 230-232

Shearer DA, Boch R, Morse RA, Laigo FM (1976) Occurrence of 9-oxodec-trans-2-enoic acid in queens of *Apis dorsata, Apis cerana* and *Apis mellifera.* J Insect Physiol 16: 1437-1441

Sheppard WS, McPheron BA (1986) Genetic variation in honey bees from an area of racial hybridization in western Czechoslovakia. Apidologie 17: 21-32

Sherebkin MW (1976) Influence of ecological conditions on catalase activity of the rectal glands. Apimondia Symp Genet Selection Reproduction 193-195

Shuel RW (1966) Einige Fragen über die Nektarsekretion. Z Bienenforsch 8: 205-209

Sidorov NG (1969) The resistance of different races of bees to experimental infection with nosema disease. Pchelovodstvo 11: 14-15

Siebold BT (1856) Wahre Parthenogenesis bei Schmetterlingen und Bienen. Leipzig

Sihag RC (1982a) Foraging behavior and pollination ecology of honeybees in relation to agricultural crops. Proc Int IUSSI Congr 9: 39

Sihag RC (1982b) Problem of wax moth *(Galleria mellonella)* infestation on Giant honeybee (*Apis dorsata* Fabr) in Haryana. Indian Bee J 44: 107-109

Silberrad REM (1976) Apiculture in Sambia. Proc Int Apic Congr 25: 342-344

Simpson H (1960) Male genitalia of *Apis* species. Nature (Lond) 185: 56

Simpson H (1970) The male genitalia of *Apis dorsata* F. (*Hym: Apidae).* Proc R Ent Soc London A 45: 169-171

Singh Y (1980) Beekeeping in Uttar Pradesh - a review. Proc Int Conf Apic Trop Climate 2: 211-226 New Delhi

Skender K (1972) Situation actuelle de l'apiculture Algerienne et ses possibilitées de development. Memoire d'ingéniorat, Alger

Skorikov AS (1929a) Beiträge zur Kenntnis der kaukasischen Honigbienenrassen. Rep Appl Entomol 4: 1-59 (Russian with German summary)

Skorikov AS (1929b) Eine neue Basis für eine Revision der Gattung *Apis* L. Rep Appl Entomol 4: 249-264 (Russian with German summary)

Smaragdova NP (1960) Influence of middle Russian hybrid nurse colonies on tongue length of South Russian bees. Pchelovodstvo 1960/9: 15-17 (Russian)

Smaragdova NP (1963) Food of worker bee larvae of *Apis mellifera mellifera* L, *Apis mellifera caucasica* Gorb and of their hybrids. Proc Int Beekeep Congr 19: 501-505

Smith F (1858) Catalogue of the Hymenopterous insects collected at Sarawak, Borneo; Mount Ophir, Malakka; and at Singapore, by A.R. Wallace. J Proc Linn Soc London Zool 2: 42-130

Smith FG (1958a) Communication and foraging range of African bees compared with that of European and Asian bees. Bee World 39: 249-252

Smith FG (1958b) Beekeeping observations in Tanganyika 1949-1957. Bee World 39: 29-36

Smith FG (1961) Races of honeybees in East Africa. Bee World 42: 255-260

Sokal RR, Sneath PHA (1963) Principles of numerical taxonomy. Freeman Co, San Francisco

Sommeijer M (1983) Social mechanisms in stingless bees. Thesis, Univ Utrecht

Sonntag C, Neureuther P, Kalinke C, Münnich K (1976) Zur Palaeoklimatik der Sahara. Naturwissenschaften 63: 479

Southwick EE (1985a) Allometric relations, metabolism and heat conductance in clusters of honeybees at cool temperatures. J Comp Physiol B 156: 143-149

Southwick EE (1985b) Bee hair structure and the effect of hair on metabolism at low temperature. J Apic Res 24: 144-149

Southwick EE (1987) Cooperative metabolism in honeybees: an alternative to antifreeze and hibernation. Int Conf Afr bees and mites. Columbus OH (in press)

Southwick EE, Mugaas JN (1971) A hypothetical homeotherm: the honeybee hive. Comp Biochem Physiol 40A: 935-944

Sperlin A, Campbell R, Brosemer RW (1975) The hybridizaton of DNA from two species of honeybee. J Insect Physiol 21: 373-376

Spinola M (1806) Insectorum Liguriae species novae aut rariores, etc. Genuae I: 159

Steche W, Böttcher FK (1983) Die Nosematose. In Zander-Böttcher (ed) Die Krankheiten der Biene. Ulmer, Stuttgart, pp 659-668

Steininger F, Rögl F (1985) Paleography and palinspastic reconstruction of the Neogene of the Mediterranean and the Parathetis. In: Dixon JE, Robertson AHF (ed) The geological evolution of the Eastern Mediterranean. Blackwell Sci Publ, Oxford pp 659-668

Stort AC (1974) Genetic study of aggressiveness in two subspecies of *Apis mellifera* in Brazil. J Apic Res 13: 33-38

Stort AC (1979) Comparative investigation of the morphometric characteristics of Italian, German, African and Africanized bees and the offspring of their hybrids. Proc Int Beekeep Congr 27: 336-339

Sturges AM (1928) Bee diseases. Bee World 9: 118-120

Svensson B (1984) Beekeeping in the Republic of Guiné-Bissau and the possibilities for its modernization. Sved Univ Agric Sci, Int Rural Dev Cent Arbetsrapport 17, Uppsala

Sylvester HA (1976) Allozyme variation in honeybees (*Apis mellifera* L). PhD Thesis Univ California, Davis

Sylvester HA (1982) Electrophoretic identification of Africanized honeybees. J Apic Res 21: 93-97

Tanabe Y, Tamaki Y (1986) Biochemical genetic studies on *Apis mellifera* and *Apis cerana*. Proc Int Beekeep Congr 30:

Taylor ORjr (1985) African bees: potential impact in the United States. Bull Entomol Soc Am 31: 15-24

Thakar CV (1973) A preliminary note on hiving *Apis dorsata* colonies. Bee World 54: 24-27

Thakar CV (1976) Practical aspects of bee management in India with *Apis cerana indica*. Proc Conf Apic Trop Climate: 51-59. IBRA Gerrards Cross

Thakar CV, Tonapi KV (1961) Nesting behaviour of Indian honeybees.1. Differentation of worker, queen and drone cells of the combs of *Apis dorsata* Fabr. Bee World 42: 61-62, 71

Thakar CV, Tonapi KV (1962) Nesting behaviour of Indian honeybees. 2. Nesting habits and comb cell differentiation in *Apis florea* Fabr. Indian Bee J 24: 27-31

Tirgari S (1971) Biology and behavioral characteristics of the Iranian Dwarf honeybee (*Apis florea*). Proc Int Beekeep Congr 23: 344-345

Tirgari S, Mirabzadeh EA (1987) Exploitation and protection measures of *Apis florea* in Iran. Proc Int IUSSI Congr (in press Peperny Munich)

Tokuda Y (1922) A pore of the *japonica* drone cocoon. Bee World 3: 89

Tokuda Y (1924) Studies on the honeybee, with special reference to the Japanese honeybee. Tr Sapporo Nat Hist Soc 9: 1-6

Tokuda Y (1935) Studies on the honeybee, with special reference to the Japanese honeybee. Bull Zootech Exp Sta Chiba Shi

Tokuda Y (1971) On the biological characteristics of the indigenous Japanese bees. Proc Int Beekeep Congr 23: 348-349

Topalidis N (1956) Apiculture on the Peninsula Chalkidike. Summaries Int Apic Congr 16: 87

Towne WF (1984) Acustic and visual cues in the dances of four honeybee species. Behav Ecol Sociobiol 16: 185-187

Tribe GD, Fletcher DJC (1977) Rate of development of the workers of *Apis mellifera adansonii* Latr. Proc Apimondia Symp Afr Bee 115-119

Tulloch AP (1980) Beeswax - composition and analysis. Bee World 61: 47-62

Ursu NA (1976) Impact of flower constancy and flower migration on the content of free amino acids in the food and on brood nursing. Proc Apimondia Symp Genet Selection Reproduction 208-213

Valette P (1922) L'apiculture dans les colonies. Proc Int Beekeep Congr 6: 44-68

Valli M, Leporati M, Cavicchi S (1984) Morphometric analysis of bee populations in Italy: northern part and Tyrrhenian coast. Abstr Int Congr Entomol 17: 508

Varma SK, Joshi NK (1983) Studies on the role of honeybees in the pollination of cauliflower (*Brassica oleracea* var *botrytis*). Indian Bee J 45: 52-53

Vats BR (1953) Breeding Italian bees in Kashmir. Kashmir 3: 23-24; 33

Vecchi A, Giavarini L (1938) Ricerche biometriche sull'*Apis mellifera ligustica*. Proc Int Congr Entomol 7: 15-30. Berlin

Veith HJ, Weiss J, Koeniger N (1978) A new alarm pheromone (2-decen-1-yl acetate) isolated from stings of *Apis dorsata* and *Apis florea* (*Hymenoptera, Apidae*). Experientia 34, 423

Velthuis HHW, Clement J, Morse RA, Laigo FM (1971) The ovaries of *Apis dorsata* workers and queens from the Philippines. J Apic Res 10: 63-66

Verma LR (1970) A comparative study of temperature regulation in *Apis mellifera* L and *Apis cerana* F. Am Bee J 110: 390-391

Verma LR (1984) Beekeeping in northern India. Proc. Expert Consult Bangkok: 148-154. FAO Rome

Verma LR (1985) Sac brood in *Apis cerana*. News Letters Trop Countries 6: 4, IBRA Cardiff

Verma LR, Edwards DK (1971) Metabolic acclimatization to temperature and temperature tolerance in *Apis mellifera* and *Apis cerana indica* in India. J Apic Res 10: 105-108

Verma LR, Mattu VK, Singh MP (1984) Races of the Indian honeybee in the Himalayas. Abstr Int Congr Entomol 17: 508

Verma S, Ruttner F (1983) Cytological analysis of the thelytokous parthenogenesis in the Cape honeybee (*Apis mellifera capensis* Escholtz). Apidologie 14: 41-58

Vesterinen F (1974) Overwintering honeybees in Finland. Sci Bull Apimondia 8. Beekeeping in cold zones, 66-67

Visscher PK, Seeley TD (1982) Foraging strategy of honeybee colonies in a temperate deciduous forest. Ecology 63: 1790-1801

Viswanathan H (1950) Temperature reading on *Apis dorsata* combs. Indian Bee J 12: 72

Vlatkovic B (1956) Beitrag zu morphologischen Untersuchungen an Bienen der Hochebene Pèster in Serbien. Abstr Int Beekeep Congr 16: 380

Vorwohl G (1968) Natürliche Diastaseschwäche der Honige von *Apis cerana* F. Z Bienenforsch 9: 232-236

Vorwohl G (1976) Honeys from tropical Africa: microscopical analysis and quality problems. Proc Apic Trop Climate 93-101. IBRA Gerrards Cross

Wafa AK, Rashad SE, Mazeed MM (1965) Biometrical studies on the Egyptian honeybee. J Apic Res 4: 161-166

Wakhle DM, Desai DB (1980) Enzymes in Indian honeys. 2nd Int Conf Apic Trop Climate 703-710. New Delhi

Walter H (1958) Klimadiagramm-Karte von Afrika. Rohrscheid, Bonn

Walter H, Breckle SV (1984) Ökologie der Erde, 2. Spezielle Ökologie der tropischen und subtropischen Zonen. Fischer, Stuttgart

Watanabe H (1921) Beekeeping in Japan. Bee World 2: 111

Weaver N, Weaver E (1981) Beekeeping with the stingless bee *Melipona Beecheii* by the Yucatanean Maya. Bee World 62: 7-19

Weiss J (1978) Vergleichende Morphologie des Stachelapparates bei den vier *Apis*-Arten (*Hymenoptera: Apidae*). Apidologie 9: 19-32

Wheeler WW (1923) Social life among the insects. Harcourt, Brace & Co

Whitcombe RP (1982) Experiments with a hive for Little Bees: some observations on manipulating colonies of *Apis florea* in Oman (Part 1). Indian Bee J 44: 57-63

Whitcombe RP (1984a) *Apis florea*. Thesis Univ Durham

Whitcombe RP (1984b) The Bedouin Bee. Aramco World News 35/21: 34-41

Wille A (1979) Phylogeny and relationships among genera and subgenera of the stingless bees (*Meliponinae*) of the world. Rev Biol Trop 27: 241-277

Wille A (1983) Biology of the stingless bees. Ann Rev Entomol 28: 41-64

Wille H (1979) Promoting bee culture in Democratic Republic of the Sudan Rep. Bee Sec Swiss Fed Inst Dairy Res Liebefeld

Williams MAJ, Faure H (1980) The Sahara and the Nile. Balkema, Rotterdam

Williams PH (1985) A preliminary cladistic investigation of relationships among the bumble bees (*Hymenoptera, Apidae*). System Entomol 10: 239–255

Williamson GP (1981) Paleontological documentation of speciation in cenozoic molluscs from Turkana Basin. Nature (Lon) 293: 437–443

Winogradowa BM (1976) Selection of the Caucasian honeybees. Apimondia Symp Gen Selection Reproduction: 246–249

Winston ML, Michener Ch (1977) Dual origin of highly social behavior among bees. Proc Nat Acad Sci USA 74: 1135–1137

Winston ML, Taylor OR, Otis GW (1983) Some differences between temperate and tropical African and South American honeybees. Bee World 64: 12–21

Woyke J (1973) Instrumental insemination of *Apis cerana indica* queens. J Apic Res 12: 151–158

Woyke J (1975a) The eggs in the comb cells of three honeybee species in India. Proc Int Beekeep Congr 25: 340

Woyke J (1975b) Natural and instrumental insemination of *Apis cerana indica* in India. J apic Res 14: 153–159

Woyke J (1976a) Brood-rearing efficiency and absconding in Indian honeybees. J Apic Res 15: 133–143

Woyke J (1976b) The density of the bristles covering the wing as discrimination value between African and other races of bees. Proc Apimondia Symp Afr Bees 15–24. Pretoria

Woyke J (1979) Sex determination in *Apis cerana indica*. J Apic Res 18: 122–127

Woyke J (1980) Evidence and action of cannibalism substance in *Apis cerana indica*. J Apic Res 19: 6–16

Woyke J (1984) Survival and prophylactic control of *Tropilaelaps clareae* infesting *Apis mellifera* colonies in Afghanistan. Apidologie 15: 421–434

Wu Yanru, Kuang Bangyu (1986) A study of genus *Micrapis* (Apidae). Zool. Res. 7: 99–102

Zander E (1941 Beiträge zur Herkunftsbestimmung bei Honig. Der Mumienhonig als pflanzen-geografische Urkunde. Loth Liedloff Michaelis, Leipzig

Zander E, Weiss K (1964) Das Leben der Biene. Ulmer, Stuttgart

Zeuner FE, Manning FJ, Morris SF (1976) A monograph on fossil bees (*Hymenoptera: Apoidea*). Bull Brit Mus Nat Hist 27: 149–268

Zymbragoudakis Ch (1979) Bees and beekeeping in Crete. Proc Int Beekeep Congr 27: 348–355

Subject Index

Numbers in *italics* indicate the page upon which the species or type in question is dealt with in most detail.